TEST Y[illegible]
REVIS[illegible]
'H' GRADE
MATHEMATICS

by

P.W. WESTWOOD, B.Sc., Dip.Ed.
Principal Teacher of Mathematics,
Kirkcaldy High School.

and

C.M. STEWART, M.A., Dip.Ed.

ISBN 0 7169 3143 5

ROBERT GIBSON · Publisher
17 Fitzroy Place, Glasgow, G3 7SF.

PREPARE TO PASS

Take Our Advice

Candidates sitting S.C.E. Revised Higher Grade Mathematics will find in this booklet the means of systematic revision for the examination.

The book, worked through methodically at the rate of ten questions each evening five nights per week, revises the whole syllabus in the last four months before the examination. It is designed to reinforce the basic knowledge and skills necessary for mathematical problem-solving at the Higher Grade. The work for the last three weeks is concentrated on facts and formulae and can be done much more quickly than the earlier sets, as we would then expect candidates to be devoting most of their time to working through previous S.C.E. Higher Mathematics examination papers. [Perhaps sets 75 to 77 should be looked at sooner.]

You should record your scores as you work through the questions, keeping an up-to-date record on the record sheet provided.

It cannot be over-emphasised that a little revision done regularly and continuously is far more effective than a massive last-minute effort.

Used properly, this book may make the difference between passing and failing, or between obtaining the grade you require and one which is not good enough.

Ask your teacher, of course, about any question which you cannot solve by yourself.

C.M.S. & P.W.W.

SET 1

1. Subtract $(3x^2 + 5x - 8)$ from $(4x^2 - 6x - 3)$.
2. Calculate the length of XY where X is the point $(1, -1)$ and Y is $(5, 3)$, leaving your answer as a surd.
3. If $f(x) = 4x^4 - 2x^2$, calculate $f(2) - f(-2)$.
4. Solve the equation $x^2 - 9 = 2x + 1$
5. Show that the points A$(-3, -9)$, B$(2, 1)$ and C$(5, 7)$ are collinear.
6. P$(-2, -3)$ and Q$(6, -3)$ are opposite ends of a diameter of a circle. Calculate the centre and radius.
7. Simplify $\sin\left(\frac{5\pi}{2} - x\right)$
8. Solve, for $0 \leqslant x \leqslant 360$, $3 \sin x° = \sqrt{3} \cos x°$ without using your calculator.
9. Find the value of the derivative of the function f defined by $f(x) = 2x^2 - 3$ at the point where $x = 5$.
10. Find $f(x)$ given that $f'(x) = 4x^3 - 2x + 1$.

SET 2

1. Expand $(x + 3)(x - 1)(x + 2)$
2. Under the transformation T, the point (x, y) is mapped on to the point $(2x - 1, 2y + 3)$. Find the image of A$(5, 2)$ under T.
3. Calculate the remainder when $x^3 - 11x + 10$ is divided by $x + 3$.
4. List the solution set of $9 > x^2$ where x is an integer.
5. Prove that the lines $y = 6$, $4y - 3x = 0$ and $x = 8$ are concurrent.
6. A is $(3, 1)$, B is $(10, -3)$ and C is $(11, 5)$. ABCD is a parallelogram. Find the co-ordinates of D.
7. Simplify $\sin(90° - A) \,.\, \tan(360° - A) \,.\, \sin(180° + A)$.
8. Find the exact value of $\sin\frac{2\pi}{3} + \sin\frac{7\pi}{3}$.
9. Calculate the gradient of the tangent to the curve $y = 9 - x^2$ at the point $(3, 0)$.
10. Differentiate $(x^2 + 4)^4$.

SET 3

1. Evaluate $\dfrac{a^2}{1+a}$ when $a = \sqrt{3} - 1$
2. Prove that triangle ABC is right angled and isosceles where A is (–3, 8), B is (–1, 2), C is (5, 4).
3. A sequence is defined by $u_n = \frac{1}{2}n(n+2)$. Find u_2 and u_7.
4. Arrange as a quadratic equation in x: $(x+c)^2 + x^2 = 9x$.
5. Find the equation of the straight line through the points (0, –3) and (3, 2).
6. Find in its general form the equation of the circle with centre (3, 1) and radius 4.
7. By expressing 75° as (45° + 30°) calculate the exact value of cos 75°.
8. Find the solution set of the equations

   (i) $\cos^2 x = \dfrac{3}{4}$ (ii) $2 \sin x = 3, 0 \leqslant x \leqslant 2\pi$.
9. $f(x) = (x^2 + 3)^2$. Find $f'(x)$.
10. Find $\int (x+2)^2 \, dx$.

SET 4

1. *(a)* Given $v = u + rt$, express r in terms of v, u and t.

   (b) Make F the subject of the formula $C = \dfrac{5}{9}(F - 32)$.

   (c) Change the subject of the formula $v^2 = u^2 - 2gs$ to g.
2. $(a, 7)$ is a point on the parabola $y = 16 - x^2$. Find a.
3. $4x^3 + 4x^2 - 11x + p$ is divisible by $x + 2$. Find p.
4. Solve the equation $x(x-4) = x$, where x is a real number.
5. A(3, –1), B(–1, 4) and C(5, 2) are three vertices of a triangle. Calculate: (i) the gradient of median AD; (ii) the gradient of altitude BE.
6. A is the point (0, 4), B(6, 0) and C(2, 2). M and N are the mid-points of AB and BC respectively. Write down the components of $\overrightarrow{MN}$.
7. Simplify $\sqrt{2}\left[\sin\left(\theta + \frac{\pi}{4}\right) - \cos\left(\theta + \frac{\pi}{4}\right)\right]$
8. Give the value of θ, $0 \leqslant \theta \leqslant \pi$, for which the maximum value of $\sin\left(2\theta - \frac{\pi}{4}\right)$ occurs.

9. State the maximum number of stationary points on the curve $y = ax^3 - 3x^2 + 4x - 5$.

10. Find the derivative of $\sqrt{x^2 + 1}$.

SET 5

1. Solve the inequality $2(3x - 1) + 3 > 7 - 3(3x - 2)$

2. Find the image of P(a, b) under
   (i) reflection in the x-axis;
   (ii) 180° rotation about O;
   (iii) 90° rotation about O clockwise;
   (iv) reflection in the line $y = x$.

3. $f(x) = \dfrac{2^x}{2^{2x} + 1}$. Evaluate $f(-1)$.

4. Find the solution set of the inequality $x^2 - 5x - 24 < 0$, where x is a real number.

5. A is (3, 2), B is (2, 3), C is (–2, –1) and D is (a, 5). Calculate the value of a if AB is parallel to CD.

6. P(–4, 3) is a point on the circle $x^2 + y^2 = 25$. Calculate the gradient of the tangent at P.

7. Simplify $\sin(A - B) + \sin(A + B)$.

8. ABCD, EFGH is a cube of side 2 cm. Calculate the angle between AG and the base ABCD.

9. Differentiate $u^{1/2} - 3u^{3/2}$ with respect to u.

10. If $f(x) = x^2$ find $f'(x)$ and $\int f(x)\,dx$.

SET 6

1. Factorise fully:
   (i) $2x^2 - 8xy$ (ii) $x^2 - 5x - 6$ (iii) $x^3 - x^2 - 2x$.

2. Write down the exact value of cos 315°.

3. Find the remainder on dividing $1 - x^3$ by $x - \frac{1}{2}$.

4. Simplify (i) $\left(\dfrac{1}{625}\right)^{-1/4}$; (ii) $27^{2/3} \times 243^{-2/5}$

5. Calculate the angle between the line AB and the positive direction of the x-axis given that A is (–1, 2) and B is (2, 6).

6. If $\overrightarrow{PQ} = k\overrightarrow{QR}$ for some number k, what can be said about P, Q and R?

7. State the maximum and minimum values of $4\cos x + 1$.

8. Arc PQ of a circle centre O radius 4 cm is 6 cm long. Calculate in radians the size of angle POQ.

9. For what value of x is there a stationary point on the curve $y = (1 - x)(x + 3)$?

10. Find $f'(x)$ if $f(x) = (x^3 + 5)^{3/2}$

SET 7

1. Solve the system of equations $2x + 3y = -5;\quad 5x - 2y = 16$

2. If $\cos x = -\frac{1}{3}$ find the **exact** value of $\sin^2 x$.

3. A sequence is defined by $u_n = \frac{1}{2}n(n + 2)$. Find n when $u_n = 24$.

4. If set X has 25 members, set Y 20 members and $X \cap Y$ has 10 members how many members are in $X \cup Y$?

5. Find the gradient of the perpendicular bisector of QR where Q is (2, –3) and R is (4, 7).

6. AB is a diameter of the circle centre (1, –3). If A is (–2, –5) find the co-ordinates of B.

7. Find x if $\sin 55° \cos x° - \cos 55° \sin x° = \frac{1}{2}, 0 \leqslant x \leqslant 180$.

8. If $\tan X\hat{Y}Z = \frac{3}{4}$, calculate the area of $\triangle XYZ$ without using your calculator.

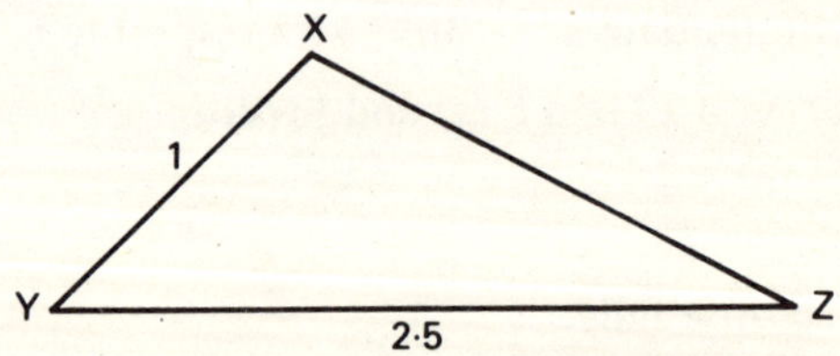

9. If $f(x) = (x + 4)^2$ find $f'(x)$.

10. $\frac{dy}{dx} = 2x + 1$ and $y = 3$ when $x = 1$. Find an expression for y.

SET 8

1. Simplify $\dfrac{2x^2 - 7x + 3}{4x - 2}, x \neq \frac{1}{2}$

   [Note: If $x = \frac{1}{2}$, then $4x - 2 = 0$ and the given fraction is not finite.]

2. Sketch the graph of $\sin x°$ for $0 \leqslant x \leqslant 360$.

3. If $x^2 + ax + 9 = (x + p)^2$ find a and p (both >0).

4. Simplify, without a calculating aid:

   (i) $\sqrt{75} - \sqrt{48}$ (ii) $\dfrac{\sqrt{27} + \sqrt{3}}{\sqrt{3}}$

5. The points (–6, 1), (2, y) and (–2, –3) are collinear. Calculate y.

6. The position vectors of P and Q are $\underline{p} = \begin{pmatrix} -2 \\ 1 \end{pmatrix}$ and $\underline{q} = \begin{pmatrix} 3 \\ -5 \end{pmatrix}$. Calculate $|\overrightarrow{PQ}|$

7. Find the value of $(3 \sin x - \cos x)^2 + (\sin x + 3 \cos x)^2$.

8. Convert 63·9° to radians.

9. Calculate the gradient of the curve $y = 5x^3 - 10x$ at the point (1, –5).

10. Differentiate $\dfrac{1}{1-x}$

SET 9

1. If $x \star y = \dfrac{x}{y} - xy$ find $2 \star 4$.

2. If $\sin x° = \dfrac{4}{5}$, $90 < x < 180$ find the **exact** value of

   (i) $\cos x°$ (ii) $\tan x°$.

3. $f(x) = x + 1$, $g(x) = 3x$ and $h(x) = x^2$. Evaluate $h\{g[f(2)]\}$.

4. Solve $6x^2 - 11x - 10 \geqslant 0$.

5. Write down the gradient of all lines perpendicular to the join of A(2, 5) and B(2, 14).

6. State whether the point (4, 6) lies **inside**, **outside** or **on** the circle $(x-1)^2 + (y+2)^2 = 25$.

7. Simplify the expression
   $\sin(A + B) . \cos(B + C) - \cos(A + B) . \sin(B + C)$.

8. ABCDE is a square right pyramid with AB = 6 cm and EC = 5 cm. Calculate the size of the angle between planes EBC and ABCD.

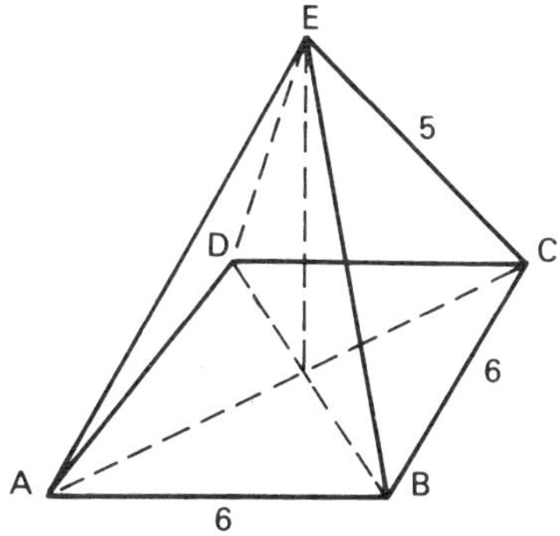

9. If $f(x) = x^3 - 2x^2 + 3$ evaluate $f'(-3)$.

10. If $f(x) = (3 - x)(3 + x)$ find (i) $f'(x)$ and (ii) $\int f(x)\,dx$.

SET 10

1. If $3^x = \frac{1}{243}$, find x.

2. The sides of a triangle are 5, 6 and 7 units respectively. Calculate the size of the largest angle in the triangle.

3. State the degree of the polynomial $x - 2x^2 + x^3 - x^4$ and give the coefficient of x^2.

4. P and Q are two sets where $P \subset Q$. Simplify
   (i) $P \cap Q$ (ii) $P \cup Q$ (iii) $P \cap P'$.

5. Find the equation of the image of the line $2x + y - 4 = 0$ under reflection in the line $y = -1$.

6. $\overrightarrow{PQ} = \binom{2}{3}$ and $\overrightarrow{QR} = \binom{a}{b}$. If P, Q and R are collinear find a relation connecting a and b.

7. Two angles of a triangle are $\frac{\pi}{4}$ and $\frac{3\pi}{5}$. Find the third angle in degrees.

8. Solve the equation $\tan^2 2x° = 1$, given $0 \leqslant x < 180$ without using your calculator.

9. Find the stationary point on the curve $y = x^3$ and determine its nature.

10. Find the derivative of $(3x^2 - 5)^{2/3}$.

SET 11

1. p varies directly as the square of q. If $q = 4$ when $p = 8$ calculate p when $q = -6$.

2. Calculate the area bounded by the lines $y = 6$, $y = x$ and $y = -x$.

3. State the 10th term of the sequence
   (i) $-1, 3, 7, \ldots\ldots\ldots\ldots$ (ii) $2, 6, 18, \ldots\ldots\ldots\ldots$

4. State the condition for the equation $x^2 + 2bx + a^2 = 0$ to have equal roots.

5. Find the equation of the line passing through A(2, –1) perpendicular to the line $2x - 3y = 5$.

6. Find the equation of the circle $x^2 + y^2 = 9$ under the translation $\binom{0}{4}$.

7. Simplify
   (i) $\cos 45° \cos 20° - \sin 45° \sin 20°$
   (ii) $\sin 90° \cos 45° + \cos 90° \sin 45°$
   (iii) $\sin 40° \sin 30° - \cos 40° \cos 30°$

8. ABCD, EFGH is a cube. Calculate the size of the angle between the base ABCD and the line DG.

9. Find the derivative of $3x^2(2 - 5x)$.

10. Find $f(x)$ given that $f'(x) = \frac{1}{x^2}, x \neq 0$.

SET 12

1. Simplify $\frac{15}{\sqrt{3}} - \sqrt{27}$ without using a calculating aid.

2. Write down the co-ordinates of the mid-point of AB where A is (–1, 2) and B is (3, 4).

3. Find the remainder and quotient when $4x^3 - 2x^2 + 4x - 5$ is divided by $x - 2$.

4. Find x if $(x^{1/3} + 1)^{1/2} = 2$.

5. Find the equation of the line through A(0, 2) which makes an angle of 135° with the x-axis.

6. A and B are two points; M is the mid-point of AB. Write down an expression for
   (i) $\overrightarrow{AB}$ in terms of $\underline{a}$ and $\underline{b}$ (ii) $\underline{m}$ in terms of $\underline{a}$ and $\underline{b}$.

7. Evaluate $\cos 450° - 2 \sin 540°$ without using your calculator.

8. In triangle ABC, AB = $\sqrt{2}$ cm, AC = $\sqrt{3}$ cm and BC = 3 cm. Calculate the value of $\cos A$.

9. Find the range of values of x for which the function f defined by $f(x) = x^2 - 4x + 3$ is increasing.

10. Find $f'(x)$ when $f(x) = \frac{1}{1 + x^2}$

SET 13

1. Simplify $(x - 3)^2 - (x + 3)^2$

2. Calculate the area of triangle PQR in which PQ = 4 cm, QR = 5 cm and PQR = 50°.

3. $f(x) = x^2 + 2x$. Find $f(x + 3)$.

4. Find the solution set of the inequality $x^2 - 3x + 2 > 0$, where $x \in R$, the set of real numbers.
5. A, B and C are (2, 0), (3, –2) and (4, 1) respectively. Prove that $\triangle$ABC is right-angled.
6. State the co-ordinates of the centre and the length of the radius of the circle $x^2 + y^2 - 4x + 2y - 11 = 0$.
7. If $\tan\left(x + \frac{\pi}{4}\right) = 2$ find, without a calculating aid, the value of tan x.

   $\left[\text{Hint: use } \tan(\alpha + \beta) = \frac{\tan\alpha + \tan\beta}{1 - \tan\alpha\tan\beta}\right]$
8. ABCD, EFGH is a cuboid with AB = 3 cm, BC = 4 cm and AE = 2 cm. Calculate the angle between AG and the base ABCD.
9. (i) $f(x) = 3 - 4x - x^2$; find $f'(-1)$.
   (ii) $f(z) = 3 - 4z - z^2$; find $f'(-1)$.
   (iii) $g(s) = 3 - 4s - s^2$; find $g'(-1)$.
10. Integrate $(3x - 1)^2$ with respect to x.

SET 14

1. Simplify (i) $8^{-2/3}$; (ii) $\frac{(2a)^3 \cdot (4a^2)^{-1}}{\sqrt{a^3} \cdot \sqrt{a^5}}$
2. The distance between the points (7, –14) and (a, 26) is 41 units. Calculate the possible values of a.
3. The remainder on dividing $x^3 + kx + 7$ by $x + 2$ is 3. Find k.
4. Simplify $\left(x - \frac{1}{x}\right)\left(x^2 + 1 + \frac{1}{x^2}\right)$, $x \neq 0$.
5. The lines $ay = 2x + 5$ and $y = 4x + 1$ are parallel. Find the value of a.
6. $\underline{a} = \binom{2}{3}$, $\underline{b} = \binom{4}{7}$ and $\underline{c} = \binom{8}{11}$
   Find m and n such that $\underline{c} = m\underline{a} + n\underline{b}$.
7. Simplify
   (i) $\sin^3 x + \sin x \cos^2 x$ (ii) $\sin^2 A + \sin^2(90° - A)$.
8. State the maximum and minimum values of $5 - \sin x$.
9. Calculate the gradient of the tangent to the curve $y = 2x^2 - 1$ at the point where $x = 0$.
10. Differentiate $\sqrt{(x + 1)(x - 1)}$

SET 15

1. Simplify $\frac{2x^2 + 5x - 3}{2x - 1}$
2. State the period of sin x and of tan x.

3. A sequence is defined by $u_n = n^2 - n + 1$.
Find (i) u_{10} (ii) n when $u_n = 73$.

4. How many subsets does $\{p, q, r\}$ have?

5. Prove that the line joining A(–2, 1) to B(1, 3) passes through C(7, 7).

6. State fully the locus of the mid-points of chords of length 8 units in a circle of radius 5 units.

7. Simplify
(i) $\cos 7x \cos 5x + \sin 7x \sin 5x$
(ii) $\cos^2 5x + \sin^2 5x$.

8. ABCD, EFGH is a cube. Calculate the size of the angle between
(i) AF and plane EHDA (ii) AG and plane CGHD.

9. Differentiate $\dfrac{1}{u^{1/2}}$ with respect to u.

10. Evaluate $\displaystyle\int_{-1}^{1} x^4\,dx$.

SET 16

1. If $a = 2$, $b = -8$ and $c = -7$ calculate $\sqrt{b^2 - 4ac}$.

2. State the general equation which describes
(i) the set of lines with gradient 3
(ii) the set of lines passing through (0, 5).

3. Express $x^2 + 6x + 8$ in the form $(x + a)^2 + b$.

4. Solve the equations (i) $2^{(x^2)} = 16$ (ii) $3^{-x} = 2^5 - 5$, where $x \in R$.

5. State the gradient of the line $\dfrac{x}{4} + \dfrac{y}{7} = 2$.

6. $\underline{u} = \begin{pmatrix} 1 \\ 0 \\ 1 \end{pmatrix}$, $\underline{v} = \begin{pmatrix} 2 \\ -4 \\ 0 \end{pmatrix}$ and $3(\underline{x} + \underline{u}) = 2\underline{v} + \underline{x}$.
Calculate the components of $\underline{x}$.

7. Copy and complete:
(i) $\cos(-A) =$ (iii) $\tan(-A) =$
(ii) $\sin(-A) =$ (iv) $\cos(270° + A) =$

8. Express $\dfrac{1 - \tan^2 A}{1 + \tan^2 A}$ in terms of $\cos A$.

9. Find the stationary point on the curve $y = x^2 - 2x$ and state its nature.

10. Find the derivative of $\dfrac{1}{x^2 - 1}$

SET 17

1. If y varies directly as x and inversely as the cube of z and $y = 1\frac{1}{2}$ when $x = 18$ and $z = 2$, find y when $x = 81$ and $z = 3$.

2. In triangle ABC, $\sin \hat{A} = 0{\cdot}8$, $\sin \hat{B} = 0{\cdot}6$ and $a = 6$ cm. Calculate b.

3. If $h(x) = (2x - 1)^2$ find $h(x + 1)$.

4. How many real roots has the equation $(x - 1)(x^2 - x + 7) = 0$?

5. A is (2, 3) and B is (–1, 6). P divides AB in the ratio 2 : 1. Calculate the co-ordinates of P.

6. The circle $x^2 + y^2 - 4x + 2y - 11 = 0$ maps on to circle $x^2 + y^2 = 16$ under a translation. Describe it.

7. Solve for $0 \leqslant x \leqslant 360$ $\quad \sin x° \cos 30° + \cos x° \sin 30° = \dfrac{\sqrt{3}}{2}$

8. V, ABCD is a right rectangular pyramid with AB = 8 cm, BC = 6 cm and height 5 cm. Calculate the angle between AV and the base ABCD.

9. If $f(x) = 2x^{3/2}$ find $f'(4)$.

10. Evaluate $\displaystyle\int_{-1}^{1} x(x^2 - 2)\,dx$.

SET 18

1. Find the solution set of the equation $x^2 - 5x = 36$, where $x \in R$.

2. State the maximum and minimum values of $\cos x$.

3. Factorise fully $x^3 - 7x + 6$.

4. On the same diagram sketch the graphs of $f(x) = x^2 - 2$ and $g(x) = 6 - x^2$. Write down the co-ordinates of the common points.

5. Find the equation of the line passing through A(2, 4) and making an angle of 45° with the x-axis.

6. If $\overrightarrow{AB} = \overrightarrow{DC}$ what kind of figure is ABCD?

7. Evaluate $\cos \dfrac{3\pi}{2} + \sin \dfrac{3\pi}{2}$ without using your calculator.

8. Simplify $\dfrac{\cos x \,.\, \sin(180° - x)}{\cos(90° - x) \,.\, \cos(180° + x)}$

9. State the range of values of x for which the function $f(x) = x^3$ is decreasing.

10. $f(x) = \dfrac{1}{\sqrt{x + 1}}$. Find $f'(x)$.

SET 19

1. Evaluate $a^{2/3} . b^{-1/4}$ when $a = 27$ and $b = 16$.
2. The line $ax + y + 1 = 0$ passes through the point of intersection of line $x = 2$ and the x-axis. Find a.
3. The sum of the first n terms of a series is $10n + 4$. Find the sum of the next n terms.
4. State the values of k for which the equation $k^2x^2 + 2x + 1 = 0$ has equal roots.
5. Show in a sketch the position of point C which divides AB externally in ratio 3:1.
6. $(h, -1)$ lies on circle $x^2 + y^2 = 10$. Find h.
7. If $\sin x° = \frac{4}{5}$ and $0 < x < 90$ calculate the exact value of $\sin 2x°$.
8. ABCD, PQRS is a cube. Calculate the size of the angle between the planes ABCD and ABRS.
9. Differentiate with respect to u: (i) $\sqrt{u}$ (ii) $\frac{1}{u^2}$
10. $\frac{ds}{dt} = 3t^2 - 1$. Find an expression for s given that $s = 3$ when $t = 2$.

SET 20

1. Simplify $\left(\sqrt{x} + \frac{1}{\sqrt{x}}\right)^2$
2. Write down the mid-point of AB where A is (5, –6) and B is (9, 8).
3. Find the value of k if $(x + 3)$ is a factor of $2x^3 - 5x^2 + kx + 18$.
4. $P = \{x: 2 \leqslant x < 4, x \in R\}$ and $Q = \{x: 3 < x \leqslant 5, x \in R\}$. Express $P' \cap Q$ in this set-builder notation.
5. Find the equation of the straight line through the origin perpendicular to $2x + 3y = 5$.
6. For what value of x are the vectors $\begin{pmatrix} -2 \\ 3 \\ 6 \end{pmatrix}$ and $\begin{pmatrix} x \\ -9 \\ -18 \end{pmatrix}$ parallel?
7. Place the following in ascending order of magnitude without using your calculator.

   $\cos \frac{2\pi}{3}$, $\sin \frac{\pi}{2}$, $\cos \pi$, $\sin \frac{\pi}{6}$, $\tan \frac{\pi}{3}$

8. Simplify $\sin^2 2A - \sin^2 A + \cos^2 2A - \cos^2 A$.

9. Calculate the angle at which the tangent to the curve $y = x^3 + x + 2$ at the point A(0, 2) is inclined to the positive direction of the x-axis.

10. Differentiate $\sin 2x$.

SET 21

1. Evaluate $c(c^2 - 5)$ when $c = 1 + \sqrt{2}$

2. Find the solution set of the equation
$$3 \cos x° + 2 = 0, 90 \leqslant x \leqslant 270.$$

3. $f : x \rightarrow x^2 - 1$ and $g : x \rightarrow x + 3$ are mappings on the set R. Find formulae for $f(g(x))$ and $g(f(x))$.

4. Write $(x + p)^2 = 3(x - q) - x^2$ as a quadratic equation in x in standard form.

5. The line through A(2, 1) and B(p, q) is parallel to the y-axis. What can be said about p or q?

6. P(2, –3) and Q(6, 7) are two diametrically opposite points on a circle. State its centre and radius.

7. Simplify
(i) $2\cos^2 A - \cos 2A - 1$ (ii) $\cos^2 3A + \sin^2 3A$.

8. E,ABCD is a square right pyramid with AB = 8 and EC = 10.
Calculate the angle between EC and the plane ABCD.

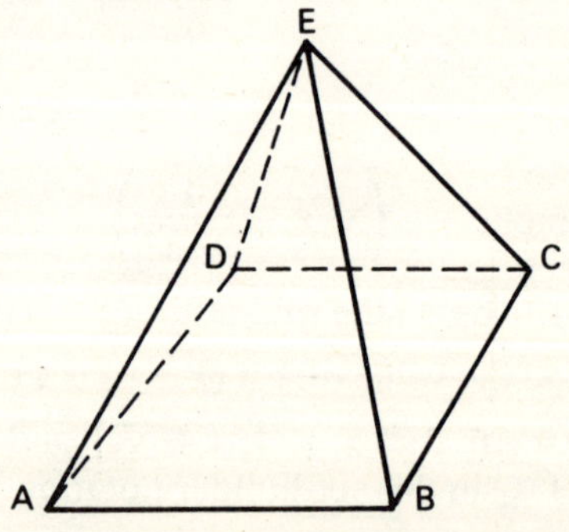

9. If $f(x) = x\sqrt{x}, x > 0$, find $f'(x)$.

10. Evaluate $\displaystyle\int_0^4 (x^{3/2} + x^{1/2})\, dx$.

SET 22

1. Solve the equation $x^2 + x = (x + 1)^2$

2. When $x = 30°$ calculate the value of the expression
$$\cos x \cos 3x + \sin 5x.$$

3. Factorise fully $x^3 + 4x^2 - 7x - 10$

4. Sketch the graph of $y = 10 - 3x - x^2$ stating the equation of the axis of symmetry.

5. The line $y = 4 - 2x$ cuts the x-axis at P and the y-axis at Q. Calculate the area of triangle OPQ.

6. O(0, 0), P(6, –4) and Q(12, 5) are vertices of a quadrilateral OPQR.
   (i) Given that $\overrightarrow{OR} = \overrightarrow{PQ}$ find the co-ordinates of R.
   (ii) Show that OQ = PR.

7. By writing 105 = 60 + 45 find the **exact** surd value of sin 105°.

8. Evaluate $(2 \sin y - 5 \cos y)^2 + (5 \sin y + 2 \cos y)^2$

9. State the range of values of x for which the function $f(x) = 3x^2 - 6x + 5$ is decreasing.

10. Find the derivative of $\cos(5x + 4)$.

SET 23

1. Give the values of x for which each of the expressions
   (i) $\dfrac{1}{x^2 - 4}$ (ii) $\dfrac{1}{x^2 - 5x + 6}$ is undefined.

2. Find the equation of the line through the point (0, 5) which is parallel to the line $3x - y = 7$.

3. A formula for the sum of the first n terms of a sequence is $S_n = 1 + \dfrac{1}{n}$
   Find the first three terms of the sequence.

4. $A = \{p, q, r, s, t, w\}$, $B = \{p, q, t\}$, $C = \{r, u, v\}$. List $A \cap (B \cup C)$.

5. A is (1, 2) and B(2, 6). Find the angle between AB and the x-axis.

6. The centre of a circle is (8, 12). The circle intersects the x-axis at (3, 0). Find the co-ordinates of the other point at which it intersects the x-axis.

7. If $\tan\theta = \dfrac{1}{5}$ and $\tan\phi = \dfrac{2}{3}$ calculate the exact value of $\tan(\theta + \phi)$, given that $\tan(A + B) = \dfrac{\tan A + \tan B}{1 - \tan A \, . \tan B}$

8. Express the sum of $\dfrac{\pi}{4}$, $\dfrac{5\pi}{6}$ and $\dfrac{\pi}{12}$ in degrees.

9. Find the rate of change of the function $y = 3x^2$ when $x = -2$.

10. Find the anti-derivative of $4 - 2x$.

SET 24

1. The sides of a right-angled triangle measure $(a-d)$, a and $(a+d)$, where $a > d$. Find the ratio $a : d$.

2. Simplify $(1 - \sin^2 A)\tan^2 A$.

3. If $(x+k)^2 = x^2 - 3x + m$ find the values of k and m.

4. Solve $x(2-3x) - 2(3x+5) \leqslant 4 - 3x(x-1)$

5. Find the value of t if the lines $3x + 4y = 7$ and $2y - tx = 6$ are perpendicular.

6. If $\underline{x} = \begin{pmatrix} 3 \\ 4 \end{pmatrix}$ and $\underline{y} = \begin{pmatrix} -5 \\ -12 \end{pmatrix}$ calculate $\dfrac{|\underline{x} + \underline{y}|}{|\underline{x}| + |\underline{y}|}$

7. For what value of x does $17\cos(x + 25)°$ have its minimum value?

8. ABCD, EFGH is a cuboid with AB = 3 cm, BC = 8 cm and AE = 5 cm. M is the mid-point of FG. Calculate the size of the angle between AM and the base ABCD.

9. If $f'(x) = (x-a)(x-b)$ state the stationary values of $f(x)$.

10. Differentiate $\dfrac{1}{\sqrt{1-x}}$.

SET 25

1. Simplify $x^{1/2}(x^{1/2} + x^{-1/2})$

2. Find the maximum value of $5 - \cos x$.

3. The functions f and g are defined on the set of real numbers by $f(x) = 2x^2 - 4$ and $g(x) = 1 - x$. Evaluate $f(g(2))$.

4. The equation $x^2 = k(2x-5)$ has equal roots. Find k.

5. Write in the form $ax + by + c = 0$ the equation of the straight line through $\left(0, -\frac{2}{3}\right)$ with gradient $\frac{4}{3}$.

6. Two concentric circles of radii 5 cm and 13 cm are drawn. A chord of the larger circle touches the inner circle. Calculate the length of the chord.

7. Simplify as far as possible $\cos^2 A + \sin^2 A - 2\sin^2 \frac{A}{2}$.

8. KLMN, PQRS is a cube. Calculate the angle between
(i) planes KLRS and NMQP (ii) planes PLMS and PQRS.

9. Differentiate $\frac{4x^2 + 3}{x}$; $x \neq 0$.

10. If $\frac{dy}{dx} = 1$, calculate y, given that $y = 0$ when $x = 0$.

SET 26

1. Evaluate $(2\sqrt{3} + \sqrt{7})(2\sqrt{3} - \sqrt{7})$ without a calculating aid.
2. State the gradient and intercept on the y-axis of the line with equation $\frac{x}{4} + \frac{y}{5} = 1$.
3. If $x - 1$ is a factor of $x^3 + a^2x^2 + 2a$ find the value of a.
4. E = {1, 2, ... 10}, A = {odd numbers}, B = {multiples of 5}. List $A \cap B'$.
5. Find p if $(5, p)$ lies on the line $2x + 6y - 7 = 0$.
6. Calculate the length of the vector $4\underline{i} - 2\sqrt{3}\,\underline{j} + 2\sqrt{2}\,\underline{k}$.
7. For what value of x, $0 \leqslant x < 360$, does $5 \sin(x - 40)°$ have its maximum value?
8. State the co-ordinates of a maximum turning point on the curve $y = 2 \cos x - 1$, where $x \in R$.
9. $f : x \rightarrow x^2$ and $g : x \rightarrow x - 1$. Find the derivative with respect to x of
   (i) $f(g(x))$ (ii) $g(f(x))$
10. $f(x) = \sin(x^2 + 2)$. Find $f'(x)$.

SET 27

1. If $\frac{1}{u} + \frac{1}{v} = \frac{1}{f}$ $(u, v, f \neq 0)$ find f in terms of u and v.
2. By using the cosine rule in $\triangle$ABC express the ratio $b \cos A : a \cos B$ in terms of a, b and c.
3. The nth term of a sequence is defined by $u_n = u_{n-1} - u_{n-2}$ with $u_1 = 0$ and $u_2 = 1$. Write down the first six terms.
4. Solve $(x + 1)^2 < 9$
5. Find y if A is $(-7, 2)$, B$(8, y)$ and AB = 17.
6. Write down the co-ordinates of the centre of the circle
   $$3x^2 + 3y^2 - 6x - 12y - 19 = 0.$$
7. If the exact value of $\cos x$ is $\frac{1}{\sqrt{5}}$ find the exact value of $\cos 2x$.
8. ABCD, EFGH is a cuboid with AB = 5 cm, BC = 10 cm and AE = 7cm. Calculate the angle between plane AFGD and the base ABCD.

9. $f(x) = \dfrac{x+1}{\sqrt{x}}$, $x \neq 0$. Find $f'(x)$.

10. Find the equation of the graph of $y = f(x)$ where $\dfrac{dy}{dx} = 2$ and $f(2) = 0$.

SET 28

1. Simplify $\dfrac{\sqrt{x} + \sqrt{x^3}}{\sqrt{x}}$
2. The line $5x + 2y - 20 = 0$ cuts the y-axis at A and the x-axis at B. The point C is the image of A under a half-turn rotation about B. Find A, B and C.
3. Factorise fully $x^3 - 3x^2 - 4x + 12$
4. Express $2^x = y$ in logarithmic form.
5. Find the equation of the perpendicular bisector of PQ where P is $(2, -1)$ and $Q(4, -3)$.
6. V, ABCD is a square pyramid. $\underline{AB} = \underline{p}$, $\underline{AD} = \underline{q}$, $\underline{AV} = \underline{r}$. Express $\underline{VC}$ in terms of $\underline{p}$, $\underline{q}$ and $\underline{r}$.
7. If $k \cos x° = 3$ and $k \sin x° = 4$ find the values of k and x.
8. State the period of (i) $\sin 4x°$ (ii) $\sin \frac{1}{4}x°$.
9. Write down an expression for the gradient of the tangent to the curve $y = f(x)$ at the point $(a, f(a))$.
10. Find the derivative of $\cos(2x - 1)$.

SET 29

1. Find the points of intersection of the curve $y = 6 + x - x^2$ and the x-axis.
2. Factorise fully $6 \sin^2 y + \sin y - 5$.
3. If $g(x) = x^4 - x^2$ evaluate $g(2) - g(-2)$.
4. Find p if $x^2 + p = 8x$ has equal roots.
5. Find the point of intersection of the lines $2x + 3y = 7$ and $5y - 3x = 18$.
6. Two circles, centres A and B and radii r_1 and r_2 touch externally. Express AB in terms of r_1 and r_2.
7. Find the maximum value of
   (i) $\cos^2 x - \sin^2 x$ (ii) $2 \sin x \cos x$.

8. P, QRST is a right rectangular pyramid with height 10 cm. The diagonals of the base measure 10 cm. Calculate the angle between PT and the base QRST.

9. $f(x) = \left(\sqrt{x} + \frac{2}{\sqrt{x}}\right)^2, x \neq 0$. Find $f'(x)$.

10. Evaluate the definite integral $\int_{-1}^{3} \sqrt{1+t}\, dt$.

SET 30

1. Which is greater, $2^{(3^2)}$ or $(2^3)^2$?
2. The line $2x - 5y = a$ passes through the point $(3, -1)$. Find a.
3. Find the value of q if the remainder is -3 when $x^3 + 2x^2 - 11x + q$ is divided by $x + 5$.
4. Express $\log_a y = p$ as an index equation.
5. Triangle OAB has vertices O(0, 0), A(7, 5) and B(1, 11). Find the equation of
   (i) the altitude OF of $\triangle$OAB (ii) the median OQ of $\triangle$OAB.
6. A is the point $(3, -1, 2)$ and B is $(2, 5, -3)$. Evaluate $\underset{\sim}{a} \cdot \underset{\sim}{b}$
7. Express $2 \cos x° + 3 \sin x°$ in the form $k \cos(x - \alpha)°$.
8. Find the solution set of the equation $\cos (x - 16)° = -\frac{\sqrt{3}}{2}, 0 < x \leqslant 360$ without using your calculator.
9. Write down an expression for the rate of change of $f(x)$ when $x = a$.
10. Differentiate $\frac{1}{2(x-2)}$

SET 31

1. Simplify $\frac{1}{(x+h)^2} - \frac{1}{x^2}$
2. Find the solution sets of these equations without using your calculator:
   (i) $\cos x° = \sin 120°$ (ii) $\tan x° = -\sqrt{3}, 0 \leqslant x \leqslant 360$.
3. A sequence is defined by $u_n = 3.4^n$. Evaluate $\frac{u_{n+1}}{u_n}$
4. If set P has 17 members, Q 29 and P $\cup$ Q 38 how many members are in P $\cap$ Q?
5. Write down the co-ordinates of the mid-point of the line joining the points $(p + q, p - q)$ and $(p - q, q - p)$.

6. Find the gradient of the tangent to the circle $(x-1)^2 + (y-2)^2 = 13$ at the point (3, 5).

7. If α and β are acute angles and $\tan\alpha = \frac{1}{2}$, $\tan\beta = \frac{1}{3}$, calculate the **exact** size of $(\alpha + \beta)$ i.e. without using your calculator.

8. PQRS, WXYZ is a cube. Calculate the size of the angle between planes PQRS and PRYW.

9. If $f(x) = \frac{(x+3)(x-1)}{\sqrt{x}}$, $x \neq 0$, find $f'(x)$.

10. Calculate the area bounded by the line $y = 2x + 3$, the x-axis and the lines $x = 0$ and $x = 2$.

SET 32

1. Simplify $(2^{1/3} - 1)(2^{2/3} + 2^{1/3} + 1)$.
2. Find the image of the point (a, b)
   (i) under reflection in the line $x = 1$
   (ii) under reflection in the line $y = -2$
   (iii) under a positive quarter turn about O.
3. Express $7 - 2x - x^2$ in the form $p - (x + q)^2$.
4. Solve $\log_2 x = 5$.
5. P is the point (3, 4). Write down the equation of the line through P
   (i) parallel to the x-axis
   (ii) parallel to the y-axis
   (iii) through the origin.
6. Given that $\overrightarrow{AB} \cdot \overrightarrow{BC} = \frac{1}{2}$, evaluate (i) $\overrightarrow{BC} \cdot \overrightarrow{AB}$ (ii) $\overrightarrow{AB} \cdot \overrightarrow{CB}$
7. Express $2\sin x° + \sqrt{5}\cos x°$ in the form $k\sin(x + \alpha)°$.
8. Solve $2\sin x\cos x = \sin x$, $0 \leqslant x \leqslant \pi$.
9. The gradient of the tangent to the curve $y = 3x^2 + 2ax + b$ at the point (1, 6) is 4. Find a and b.
10. Find $f'(x)$ if $f(x) = (1 + \sin x)^4$.

SET 33

1. If $p \star q = p^2 + q$ evaluate $2 \star (4 \star 3)$.
2. A point $(1, h)$ lies on the line $y = 2x + 3$. Find h.

3. f and g are two functions defined by $f(x) = 1 - x^2$ and $g(x) = \cos x$. Find $f(g(x))$ in its simplest form.

4. Determine the nature of the roots of the equation $3x^2 - 4x - 2 = 0$.

5. Find k if the lines $2x - y = 5$, $3x - 2y = 9$ and $4x + ky = 1$ are concurrent.

6. State the general equation of all circles which pass through the origin.

7. Express $\cos A$ in terms of (i) $\cos \frac{A}{2}$ (ii) $\sin \frac{A}{2}$.

8. Solve $\sin 2x° = 4 \sin x°, 0 < x \leqslant 360$ without using your calculator.

9. Simplify $f(x) = \left(\sqrt[3]{x} - \frac{1}{\sqrt[3]{x}} \right)^2$. Find $f'(x)$.

10. Illustrate the integral $\int_{-1}^{2} (4 - x^2)dx$ by means of a sketch.

SET 34

1. Find the solution set of the system of equations $\left. \begin{aligned} y - x &= 0 \\ x^2 + y^2 &= 2 \end{aligned} \right\}$

2. Simplify (i) $\sin(-\theta)$ (ii) $\cos(-\theta)$.

3. For what value of k is $(x + 2)$ a factor of $9x^3 + kx^2 - x - 2$?

4. If $S = \{x : x + 2 > 0, x \in R\}$ express S' in set-builder notation.

5. Triangle PQR has vertices P(4, –1), Q(7, 11) and R(10, 5). Find the equation of the altitude PT.

6. Find a if $2\underline{i} + 3\underline{j} - \underline{k}$ and $a\underline{i} + 2\underline{j} - 3\underline{k}$ are perpendicular.

7. Find the maximum value of $\sqrt{3} \cos x° + \sin x°$.

8. Simplify $\cos^2 2A + \sin^2 2A - 2 \sin^2 A$.

9. Calculate the gradient of the tangent to the curve $y = 4x - x^2$ at the point (1, 3).

10. Differentiate $\cos^2 x$.

SET 35

1. Simplify $3a^{1/2} \times 3a^{-1/2}$

2. XY = 5, YZ = 3 and XZ = 8. What can be said about the points X, Y and Z?

3. The nth term of a sequence is defined by $u_n = \frac{n}{2}[1 - (-1)^n]$. Write down u_{15} and u_{16}.

4. Solve $(x-2)^2 - 16 \geqslant 0$.

5. A is the point $(t + u, u)$ and B is $(u - t, t)$. M is the mid-point of AB. Find the gradient of OM.

6. Give the co-ordinates of the centre of the circle
$$2x^2 + 2y^2 - 4x + 7y - 1 = 0$$

7. If $\tan x = \frac{1}{2}$, find without a calculating aid, the **exact** value of $\sin 2x$.

8. ABCD, EFGH is a cuboid with AB = 2 cm, BC = 4 cm and AE = 3 cm. Calculate the angle between BH and plane BCGF.

9. Find $\frac{d}{dx} a(x^3 - x^2)$, $a \in R$.

10. Evaluate $\int_{-\pi}^{\pi} (1 + \cos x)dx$.

SET 36

1. Sketch the graph of $y = f(x)$ where $f : x \rightarrow x^2 - 6x + 9$.

2. Calculate the **exact** value of $\sin^2 120° \cos 180° + \tan^2 135°$.

3. Factorise fully $2x^3 + 4x^2 - 10x - 12$.

4. Express $q^n = 3$ in logarithmic form.

5. Find t if the line $2tx + (1 - t)y + 1 = 0$ passes through the point (4, 5).

6. Given that $|\underline{p}| = 3$, $|\underline{q}| = 4$ and $\underline{p} \cdot \underline{q} = 9$ evaluate
(i) $\underline{p} \cdot (\underline{p} + \underline{q})$ (ii) $(\underline{p} - \underline{q}) \cdot \underline{q}$

7. Find the minimum value of $2 \cos x - 3 \sin x$.

8. This diagram shows part of the graph of a simple trigonometrical function.
State the likely function.

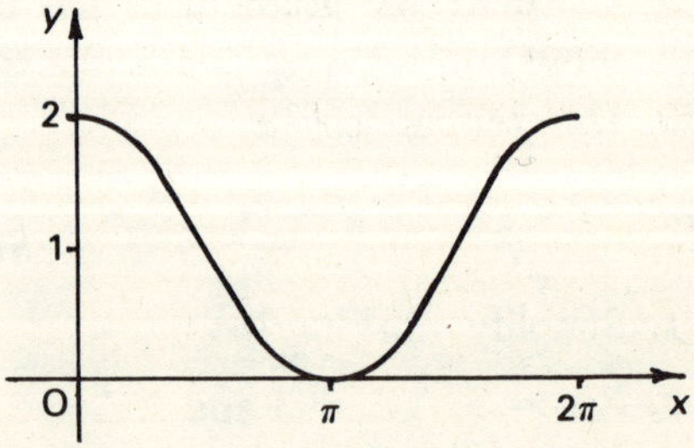

9. Calculate the rate of change of the function f defined by $f(t) = 10t - 5t^2$ when $t = 2$.

10. Find the derivative of $\sin(2x^2 + 1)$.

SET 37

1. If $f(x) = 2x^2 - 3x + 1$ simplify $f(3x) - f(2x)$.

2. Write down the co-ordinates of the mid-point of AB where A is (6, 4) and B is (–4, –6).

3. $f : x \rightarrow x + 3$ and $g : x \rightarrow x^2$. Find a formula for the function $f(g(f(x)))$.

4. Show that the roots of $5x^2 - kx - 1 = 0$ are always real.

5. Find the equation of the straight line through (0, 2)
(i) parallel to $3x - y = 7$ (ii) perpendicular to $5x - 4y = 10$.

6. Write down the equation of the circle which is the image of the circle $(x - 1)^2 + (y - 2)^2 = 9$ under reflection in the x-axis.

7. Without a calculating aid, evaluate $(\sin 63° \cos 27° + \cos 63° \sin 27°)^2$

8. ABCD, EFGH is a cube. Find the angle between planes EFGH and BDHF.

9. Evaluate $f'(1)$ where $f(z) = 4z - z^2$.

10. If $g(x) = (3x - 1)^2$ find (i) $g'(x)$ (ii) $\int g(x)dx$.

SET 38

1. Solve the equation $x^4 - 8x = 0, x \in R$.

2. Factorise fully (i) $2 \sin^2 x + 5 \sin x - 3$ (ii) $\cos^2 A + 2 \sin A - 1$.

3. A sequence is defined by the recurrence relation $u_{n+1} = 3u_n + 4$ and $u_o = -1$. Which is the first term of this sequence to exceed 100?

4. Evaluate $\log_4 64$

5. Find y if AB is perpendicular to CD where A is (4, 5), B(5, 3), C(4, –5) and D(3, y).

6. Evaluate $\underline{a} . \underline{b}$ where $\underline{a} = 2\underline{i} - 3\underline{j} + \underline{k}$ and $\underline{b} = 3\underline{i} - 2\underline{j} + 4\underline{k}$

7. Find the value of x for which $7 \cos(x + 300)°$ has its maximum value.

8. Calculate the exact value of $\sin^2 750°$ without using your calculator.

9. Find the gradient of the tangent to the curve whose equation is $s = 5t^2$ at the point where $t = -1$.

10. If $y = (x^2 + 3x - 5)^4$ find $\frac{dy}{dx}$

SET 39

1. $f(a) = 4a - 3$. Evaluate $\lim_{h \to 0} \frac{f(1 + h) - f(1)}{h}$

2. The line $y = 2x - 3$ is reflected in the x-axis. State the equation of the image.

3. Solve $x^3 + 2x^2 - x - 2 = 0$

4. If set A has 53 members, set B has 49 members and $A \cap B$ has 17 members how many members does $A \cup B$ have?

5. Triangle ABC has vertices A(–5, 2), B(1, 9) and C(5, 3). Find the equation of the median AD.

6. State the centre and radius of the circle with equation
$$x^2 + y^2 - 4x + 6y - 3 = 0.$$

7. If $\sin x° = -\frac{4}{5}$ and $270 \leqslant x \leqslant 360$, evaluate $\sin 2x°$ exactly.

8. PQRS, WXYZ is a cuboid with PQ = 3 cm, QR = 8 cm and PW = 5 cm. M is the mid-point of ZY. Calculate the angle between planes PQM and PQRS.

9. Given that $V = \pi r^2 h$, where r is constant, find the derivative of V with respect to h.

10. Sketch the graph of the function f given by $f(x) = x^3$. Given that $\int_0^1 x^3 \, dx = \frac{1}{4}$, use your graph to evaluate $\int_{-1}^0 x^3 \, dx$.

SET 40

1. Simplify $\frac{a^2 - 7a + 12}{a^2 - 9} \times \frac{a^2 + 2a - 3}{a^2 - 8a + 16}$

2. Calculate the smallest angle in the triangle with sides 2, 3 and 4 cm.

3. Express $x^4 + 2x^2 + 5$ in the form $(x^2 + a)^2 + b$ and hence state its minimum value.

4. Solve $\log_{\sqrt{3}} x = 8$

5. Prove that A(–6, 3), B(2, –5) and C(–2, –1) are collinear.

6. Simplify $\underline{x} \cdot (\underline{y} - \underline{z}) + \underline{y} \cdot (\underline{z} - \underline{x}) + \underline{z} \cdot (\underline{x} - \underline{y})$

7. Find the maximum value of $1 + 3 \cos x - 4 \sin x$.

8. This diagram shows part of the graph of a simple trigonometrical function.

State the likely function.

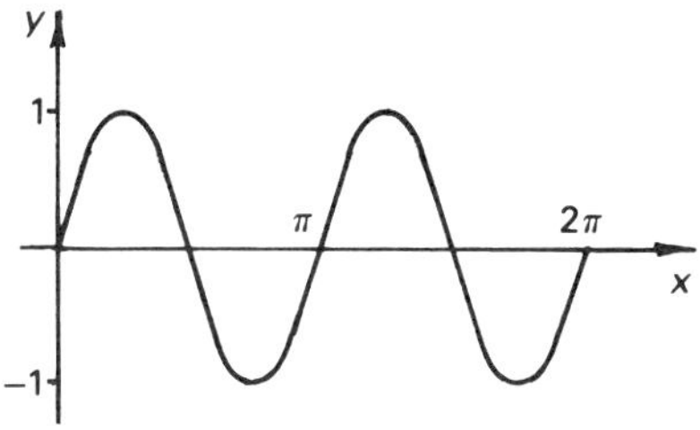

9. $y = x^3 + 1$. Show that y is never decreasing.

10. Differentiate $\cos(3x^2 - 5)$

SET 41

1. Solve the equation $(x - 1)(x - 2) = 6$, where $x \in R$.

2. Write down the equation of the line through the point (0, 3) perpendicular to the line $3x - 2y = 4$.

3. Which elements of the domain of the function $f(x) = 7 - x^2$ map to 3?

4. For what values of k does the equation $kx^2 + 2x + (2k - 1) = 0$ have equal roots?

5. A is (3, 4), B(7, 2) and C(9, 1). Find the equation of the line through A parallel to BC.

6. Find the equation of the image of the circle $x^2 + y^2 - 2x - 4y - 4 = 0$ under reflection in the y-axis.

7. If $\tan x = 2$ find the **exact** value of $\cos 2x$.

8. Rectangle ABCD is vertical and isosceles triangle ABE is horizontal. AB = 10 cm, AD = 4 cm, AE = 13 cm. Calculate the angle between planes DCE and ABCD.

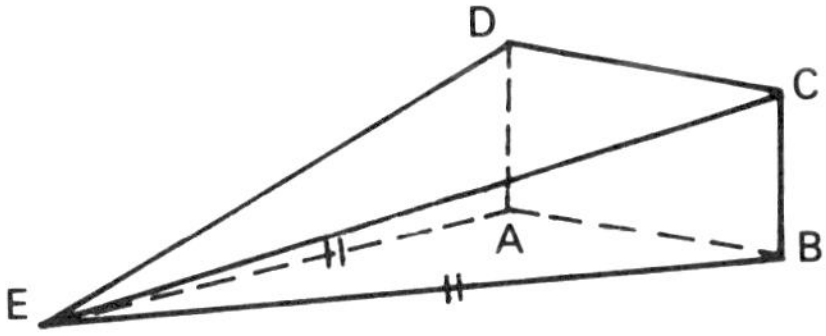

9. Differentiate with respect to α
(i) $\sin \alpha$ (ii) $\sin(2\alpha + 1)$ (iii) $\sin(2 - 3\alpha)$.

10. Evaluate $\int_{-5}^{5} x \, dx$.

SET 42

1. If $g(x) = 3x^2 + 1$, evaluate $g(x + h) - g(x)$.
2. In the circle, $A\hat{O}B = 60°$.
Calculate, correct to 1 decimal place
(i) the length of arc AB,
(ii) the area of sector AOB,
(iii) the area of $\triangle$AOB.

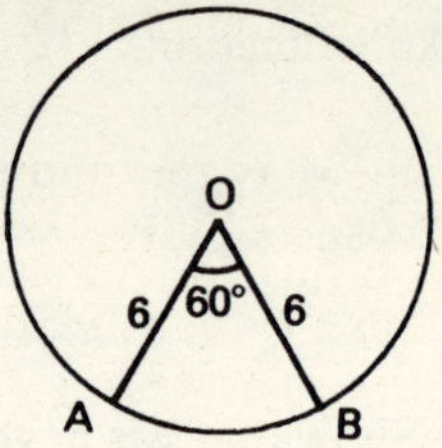

3. Factorise fully $2x^3 + 5x^2 - 23x + 10$.
4. How many subsets does the set $\{1, 2, 3, 4\}$ have?
5. Find z if PQ is perpendicular to RS where P is (4, 2), Q(6, 4), R(0, –3) and S(2, z).
6. Express cos $P\hat{Q}R$ in terms of $\overrightarrow{QP}$ and $\overrightarrow{QR}$.

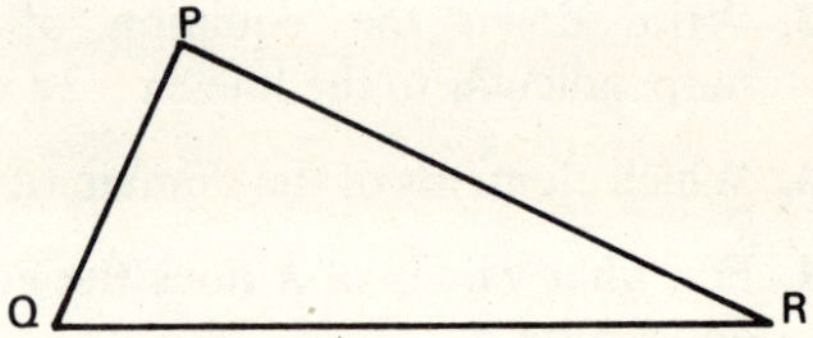

7. Find the minimum value of $4 - 5 \cos x + 12 \sin x$.
8. $f : x \rightarrow \cos^2 x - \sin^2 x, x \in R$. State the maximum value of f.
9. Find the gradient of the curve $y = x(x - 2)$ at the point (3, 3).
10. Write down the derivative of $(x^2 - 5x - 2)^{3/2}$

SET 43

1. Simplify $3x^2y^{-1/2} . (2xy^{-1/3})^{-2}$
2. A line PQ makes an angle of 40° with the positive x-axis. Line RS is perpendicular to PQ. Give the size of the angle which RS makes with the x-axis.
3. Find the first five terms of the sequence defined by $u_{n+1} = 10u_n - 9$ when $u_1 = 2$.
4. Determine the nature of the roots of the equation $(x - 2)(x - 1) = 30$.
5. The line $y = 10 - 2x$ cuts the x-axis at A and the y-axis at B. Find the co-ordinates of the mid-point of AB.

6. Find the equation of the circle which passes through the points (0, 1), (0, 5), (1, 0), (5, 0).

7. Simplify $\sin(60° + X) - \sin(60° - X)$

8. ABCD, EFGH is a cuboid with AB = 5 cm, AD = 3 cm and AE = 2 cm. M is the mid-point of BC. Calculate the angle between planes EHM and EFGH.

9. $f(x) = 2 + \sin x + \cos x$. Find $f'(x)$.

10. Illustrate by a sketch $\int_{-1}^{2} x^2\,dx$.

SET 44

1. Solve the system of equations $\left.\begin{aligned} 2x - 3y &= 3 \\ -4x + 2y &= 10 \end{aligned}\right\}$

2. Solve the equation $\sin(x - 50)° = \frac{1}{2}$, $0 \leqslant x \leqslant 360$.

3. Find p if the remainder on dividing $x^4 + px^3 - 5x + 11$ by $x - 3$ is -4.

4. Express in index form $p = \log_2 q$

5. Find the equation of the line through (2, −1) perpendicular to the line $x + 3y = 4$.

6. Calculate the cosine of the angle between the vectors $\begin{pmatrix} 2 \\ -1 \\ 2 \end{pmatrix}$ and $\begin{pmatrix} 3 \\ 6 \\ 2 \end{pmatrix}$

7. Find the value of x for which $10\sin(x - 30)°$ has its minimum value.

8. State the period of (i) $\sin^2\theta$ (ii) $\sin 2\theta$ (iii) $\sin\frac{1}{2}\theta$.

9. Calculate the rate of change of the function $f: x \to \sqrt{x}$ when $x = 4$.

10. Differentiate $\sin^3 x$.

SET 45

1. Expand $(a - b)(a + 2b - c)$

2. In $\triangle$ABC, $a = 5$, $b = 8$. State the range of possible values for c.

3. A sequence is defined by $u_n = 4n + 7$. Evaluate $u_{n+1} - u_n$.

4. Show that for all values of k the equation $k^2x^2 - 2kx + 1 = 0$ has equal roots.

5. PQRS is a kite with P(6, −2) and R(3, −5). Find the gradient of QS.

6. Find the radius of a circle which passes through (0, 1) and (0, 3) and touches the x-axis.

7. Find, without using your calculator, the solution set of $\sin x = \sin 2x$, $0 \leqslant x \leqslant 2\pi$.

8. E, ABCD is a right square pyramid with AB = 6 cm and vertical height 9 cm. Calculate the size of $D\hat{E}B$.

9. If $A = 2\pi r(r + h)$, where h is constant, find the derivative of A with respect to r.

10. $f'(x) = x(2x^2 - 1)$. Find $f(x)$.

SET 46

1. Factorise $(a-b)^2 - (b-c)^2$

2. Calculate the area of triangle ABC, given $a = 4$ mm, $b = 5$ mm and $\hat{C} = 30°$.

3. Solve $3x^3 + 4x^2 - 13x + 6 = 0$

4. Simplify $\log a + \log a^2 - \log a^3$

5. A is (2, 1), B (3, 4), C (5, 7) and D (8, –2). P and Q are the mid-points of AC and BD respectively. Find the gradient of PQ.

6. R divides EF in the ratio 3 : 2. Find the co-ordinates of R if E is (5, 0, 10) and F is (0, 10, –5).

7. Find the maximum value of $\cos x° + \sin x°$ and the value of x for which it occurs.

8. $\sin x° = \frac{1}{\sqrt{3}}$ $(0 < x < 90)$. Find the exact values of (i) $\cos x°$ (ii) $\cos 2x°$.

9. Calculate the gradient of the tangent to the curve $y = \frac{1}{3}x^3 + \frac{1}{2}x^2$ at the point where $x = -1$.

10. $f(x) = (1 - 2\cos x)^3$. Find $f'(x)$.

SET 47

1. Simplify $\frac{\sqrt{75} - \sqrt{50}}{\sqrt{12} - \sqrt{8}}$

2. Write down the equation of the line through the point (5, 6) (i) parallel to x-axis (ii) parallel to y-axis.

3. (i) Sketch the graph of the function defined by $f(x) = \begin{cases} x + 2 : x \leqslant 0 \\ 2 - x : x > 0 \end{cases}$

(ii) Explain why f does not have an inverse.

4. List $A = \{(x, y) : x + y < 2, x, y \in W\}$, $W = \{0, 1, 2, \ldots\}$

5. The line l has the equation $y = 2x + 1$. The line m is parallel to l but 4 units to the right. Find the equation of m.

6. Find the equation of the tangent to the circle $x^2 + y^2 = 25$ at the point $(-4, 3)$.

7. Simplify (i) $2\cos^2 A - 1 - \cos 2A$ (ii) $2 \sin 2A \cos 2A + \sin(-4A)$.

8. PQRS, TUVW is a cube of side 4 cm. M and N are the mid-points of UV and VW respectively.

Calculate (i) the length of PM (ii) the size of $P\hat{M}N$.

9. Find the rate of change of the function f given by $f : x \rightarrow \sin x$ at the point $\left(\frac{3\pi}{2}, -1\right)$.

10. Given that $\int_a^b f(x)\,dx = k$ write down the value of $\int_b^a f(x)\,dx$.

SET 48

1. Simplify $\dfrac{x^2 - 4}{6x - 9} \times \dfrac{2x^2 - 3x}{3x^2 + 5x - 2}$

2. State the number of elements in the solution set of the equation $2(\sin x° + 1)(3 \sin x° - 4) = 0, x \in R, 0 \leqslant x < 360$.

3. By completing the square find the least possible value of $x^2 + 6x$.

4. Evaluate $\frac{1}{2}\log_7 49$

5. Find the point of intersection of the lines $2x - y = 8$ and $3x + 2y = 5$.

6. Evaluate $(\underline{a} + \underline{b}) . (\underline{a} - \underline{b})$.

7. Find the minimum value of $\cos x° - \sin x°$ and the value of x for which it occurs.

8. Simplify $\dfrac{\sin x° - \cos(90 + x)°}{\sin x°}$

9. Find the range of values of x for which the function g defined by $g(x) = 2x^3 - 6x$ is increasing.

10. Differentiate $(x^2 + \sin x)^2$

SET 49

1. $V = \frac{\pi r^2}{3}(2r + 3h)$. Change the subject to h.
2. The line $x + 2y = 6$ cuts the x-axis at A and the y-axis at B. Find the mid-point of AB.
3. The functions f and g are defined by $f(x) = 1 - 2x^2$, $g(x) = \sin x$. Express $f(g(x))$ in its simplest form.
4. For what range of values of k does the equation $3x^2 + 4x + k = 0$ have real roots?
5. Find the equation of the line through $(3, -2)$ and $(5, 1)$.
6. Find the equation of the circle passing through the origin with radius $\sqrt{8}$ and centre on the line $y = x$.
7. If $\tan\theta = \sqrt{3}$ calculate the exact value of $\tan 2\theta$.
$\left[\tan(x + y) = \frac{\tan x + \tan y}{1 - \tan x \,.\, \tan y}\right]$
8. Simplify $\tan x^\circ + \frac{1}{\tan x^\circ}$
9. Differentiate $a^{\frac{1}{2}}(a^{\frac{1}{2}} - a^{\frac{-1}{2}})$ with respect to a.
10. Integrate with respect to x (i) $2x + 1$ (ii) $(2x + 1)^3$ (iii) $(2x + 1)^7$.

SET 50

1. r varies as the square of x and inversely as t. When $x = 4$ and $t = 2$, $r = 24$. Find r when $x = t = 1$.
2. Find the solution set of the inequality $\sin\theta > \frac{1}{2}$, where $0 \leqslant \theta \leqslant \pi$.
3. Factorise fully $x^4 - 5x^2 + 4$
4. Solve $x^{2\cdot5} = 9$.
5. Write down the gradient of the line $5x + 7y = 12$.
6. If $\underline{a}$ and $\underline{b}$ are vectors such that $|\underline{a}| = |\underline{b}|$ calculate the size of the angle between the vectors $\underline{a} + \underline{b}$ and $\underline{a} - \underline{b}$.
7. Find the minimum value of $3\cos x^\circ + 4\sin x^\circ$ and the value of x for which it occurs.
8. Sketch the graph of $\cos 2x^\circ + 1$ for $0 \leqslant x \leqslant 360$.
9. Find the gradient of the tangent to the curve $y = f(x) = \cos 3x$ at the point $\left(\frac{\pi}{2}, 0\right)$
10. Find the derivative of $\frac{1}{\sqrt{(2x + 3)(x - 2)}}$

SET 51

1. Simplify $\dfrac{a^n \times (a^{n-1})^n}{a^{n+1} \times a^{n-1}}$

2. A sector of a circle has radius 8 cm and the angle at the centre is 30°. Calculate the length of the arc of the sector in terms of π.

3. A sequence is defined by $u_n = 1 + \dfrac{2}{n^2}$. Evaluate $\lim\limits_{n \to \infty} u_n$.

4. Determine the nature of the roots of the equation $(2-x)(3-x) + 1 = 0$.

5. Which point on the line $x + 3y = 10$ is closest to the origin?

6. Find the equation of the tangent at $(7, -1)$ to the circle $x^2 + y^2 - 6x + 8y = 0$.

7. If $\alpha + \beta = \dfrac{\pi}{4}$ and $\tan \alpha = \dfrac{3}{4}$, find the **exact** value of $\tan \beta$, using
$$\tan (x + y) = \frac{\tan x + \tan y}{1 - \tan x \,.\, \tan y}$$

8. V, ABCD is a right rectangular pyramid with AB = 8 cm, AD = 6 cm, AV = 7 cm. Calculate the angle between planes VAC and VBD.

9. $f(x) = \sqrt{3x - x^2}$. Find $f'(x)$.

10. Given that $a > 1$ and $\int_1^a (2x + 3)\, dx = 24$, find a.

SET 52

1. On the same diagram sketch the graphs of $y = x^2 + 2$ and $y = (x + 2)^2$.

2. Express sin 402° as the sine of an acute angle and as the cosine of an acute angle, without using your calculator.

3. Solve $2x^3 - 19x^2 + 59x - 60 = 0$

4. Express $1 + \log_3 2$ as a single logarithm.

5. Find t if $(3t + 2, 2t - 3)$ lies on the line $2x + 3y = 6$.

6. V, ABCD is a square pyramid. The diagonals of ABCD meet at E. M is the mid-point of VE. $\underline{\text{AB}} = \underset{\sim}{p}$, $\underline{\text{AD}} = \underset{\sim}{q}$, $\underline{\text{AV}} = \underset{\sim}{r}$. Express $\underline{\text{VM}}$ in terms of $\underset{\sim}{p}$, $\underset{\sim}{q}$ and $\underset{\sim}{r}$.

7. Find the maximum value of $4 \cos x° - 3 \sin x°$ and the value of x for which it occurs.

8. Solve the equation $4 \sin^2 x° = 1, 0 \leqslant x \leqslant 360$.

9. The line $y = mx + 3$ is perpendicular to the tangent at P(1, 0) to the curve $y = 1 - x^2$. Find m.

10. $f(x) = (1 - \sin x)(1 + \sin x)$. Find $f'(x)$.

SET 53

1. Factorise fully $18x^2y - 33\,xy^2 - 30y^3$
2. The line $y - mx = 4$ is perpendicular to the line $2y + x = 3$. Find m.
3. A function g is defined by $g(x) = [g(x-1)]^2$. If $g(0) = 2$, find $g(2)$.
4. Show that if the roots of $7x^2 - px + 1 = 0$ are real then the roots of $8x^2 + 2px + 3 = 0$ are also real.
5. Find the point of intersection of the lines $2x + 4y = 10$ and $3x - 2y = 25$.
6. The circle $x^2 + y^2 - 3x + 7y - 10 = 0$ cuts the x-axis at A and B. Find the length of chord AB.
7. Find $\cos\theta$ if $\cos 2\theta + 3\cos\theta = 1$.
8. Sketch the graph of $1 - \sin x$, $0 \leqslant x \leqslant 2\pi$.
9. Find the rate of change of the function f given by $f(s) = 4s - s^2$ when $s = 1$.
10. Write down $\sin^2 u$ in terms of $\cos 2u$ and then find $\int \sin^2 u\, du$.

SET 54

1. Expand $(2x + 3)(x - 2)(3x + 5)$
2. State the period of (i) $\cos x°$ (ii) $\cos^2 x°$.
3. A sequence is defined by the recurrence relation $u_{n+1} = u_n^2 - 2n + 1$. If $u_0 = 0$ how many terms of the sequence are negative?
4. Solve $3^x = 7$
5. Write down the equation of the image of the line $3y = 4x$ under reflection in the line $y = x$.
6. Draw a counter-example to disprove the conjecture $\underset{\sim}{a} \cdot \underset{\sim}{b} = \underset{\sim}{b} \cdot \underset{\sim}{c} \Rightarrow \underset{\sim}{a} = \underset{\sim}{c}$.
7. Sketch the graph of $\cos x° - \sin x°$ for $0 \leqslant x \leqslant 360$.
8. Write down the supplement of $\frac{8\pi}{15}$ in degrees.
9. Show that the function g defined by $g(x) = x + \sin x$ is never decreasing.
10. Find the derivative of $(ax^2 + b)^n$.

SET 55

1. Establish a formula for the shaded area.

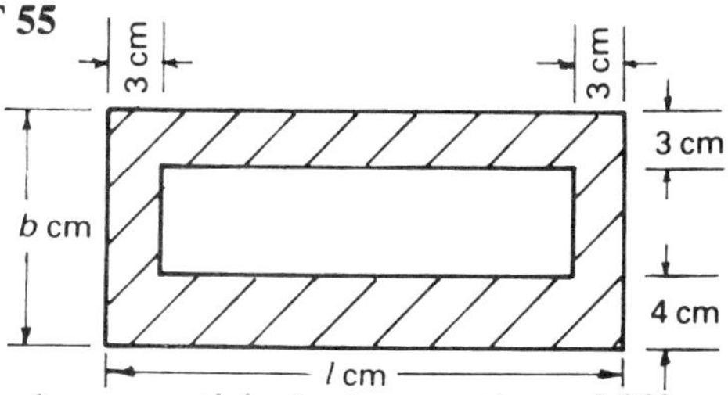

2. P′(5, 5) is the image of P(3, 1) under an anticlockwise rotation of 90° about a point Q. Find the co-ordinates of Q.
3. For the function $f(x) = \sqrt{9-x^2}$ state (i) the domain; (ii) the range.
4. List $\{(x, y) : -6 < x < -4;\ 1 \leqslant y \leqslant 3;\ x, y \in Z\}$. ($Z = \{0, \pm 1, \pm 2, \ldots\}$)
5. The line $\frac{x}{4} + \frac{y}{5} = 1$ cuts the x-axis at P and the y-axis at Q. Calculate the area of triangle OPQ.
6. The line $3x + 2y = 13$ is the tangent at P to a circle with centre O. Find
   (i) the equation of OP (ii) the co-ordinates of P
   (iii) the equation of the circle.
7. Factorise fully $\cos 2x + \cos x$.
8. Calculate the size of the angle between any two faces of a regular tetrahedron.
9. If $f(x) = \cos 3x$ find $f'\left(\frac{\pi}{2}\right)$.
10. Write down an integral which is a measure of the shaded area.

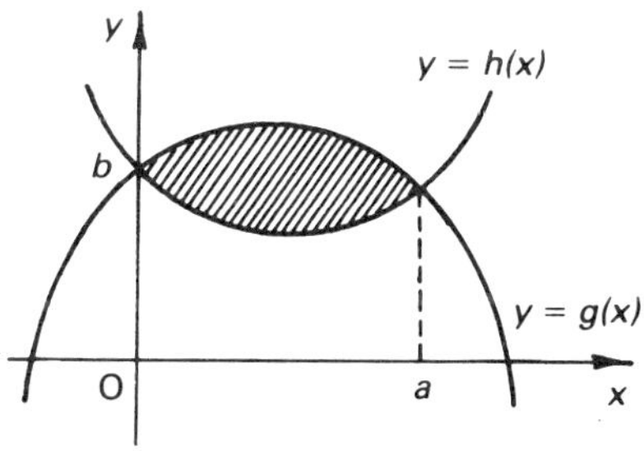

SET 56

1. Find all the values of k for which the solution set of $\begin{cases} x^2 + y^2 = 4 \\ y = k \end{cases}$ is non-empty.
2. Solve the equations
   (i) $3\cos^2 x° - \cos x° = 0$
   (ii) $\sin^2 x° - 3\sin x° + 2 = 0,\ 0 \leqslant x \leqslant 360$.

3. By completing the square find the greatest possible value of $3 + 10x - x^2$ and give the corresponding value of x.

4. Express $3 \log x - 4 \log \sqrt{y}$ as a single logarithm.

5. Calculate the distance between the points (1, 5, 7) and (0, –3, 6).

6. Find the cosine of the angle between vectors $\underline{p} = 2\underline{i} + \underline{j} - 2\underline{k}$ and $\underline{q} = 3\underline{i} - 4\underline{k}$.

7. What is the minimum value of $k = 17 + 5 \sin (x - 35)°$? For what value of x does it occur?

8. Simplify

(i) $\cos (\pi + \theta)$ (ii) $\sin \left(\frac{\pi}{2} - \theta\right)$ (iii) $\cos (2\pi - \theta)$ (iv) $\sin \left(\frac{3\pi}{2} - \theta\right)$.

9. Find the co-ordinates of the points at which the tangent to $y = x^3 + 3x^2 - 2x + 5$ is perpendicular to the line $2x - 4y = 5$.

10. Differentiate $\cos^3 2x$.

SET 57

1. State the nature of the roots of the equation $2x^2 - 5x - 7 = 0$.

2. Without using your calculator, give the **exact** value of sin 600°.

3. Two functions f and g are defined as follows:

$$f(x) = x - 1;\ g(x) = (x + 1)^2 + x + 1.$$

Find in its simplest form an expression for $g(f(x))$.

4. Express $\frac{1}{2} \log_{10}(3^2 + 4^2)$ in terms of $\log_{10} 2$.

5. P and Q are points $(2, a)$ and $(5, b)$ on the curve $xy = 20$. Calculate the gradient of PQ.

6. $\underline{i}$, $\underline{j}$ and $\underline{k}$ are unit orthogonal vectors. Evaluate (i) $\underline{i} \cdot \underline{i}$ (ii) $\underline{j} \cdot \underline{k}$.

7. If A and B are acute angles such that $\cos A = \frac{1}{7}$ and $\cos B = \frac{13}{14}$, find the **exact** values of $\sin A$ and $\sin B$.

8. For A and B in Question 7 find the **exact** value of $\cos (A - B)$.

9. $h(x) = \left(\sqrt{x} + \frac{1}{\sqrt{x}}\right)\left(\sqrt{x} - \frac{1}{\sqrt{x}}\right)$. Find $h'(x)$.

10. Show in a sketch the area given by $\int_{-2}^{2} x^3\, dx$. Hence state the value of $\int_{-2}^{2} x^3\, dx$.

SET 58

1. In a class of 25 boys, 15 play rugby, 16 play football and 4 play neither. How many play both rugby and football.

2. Factorise $\cos^2 A - 3\cos A + 2$

3. Find the remainder on dividing $13 + 7a - 5a^3$ by $a + 4$.

4. Re-arrange the equation $k = \dfrac{a^2 + 3a + 4}{a + 2}$, $a \neq -2$, as a quadratic in a.

5. A is the point $(-1, 0, 2)$ and B is $(0, 1, -2)$. Find the co-ordinates of P which divides AB externally in the ratio 2 : 1.

6. State in terms of θ and a the radius of the circle
$$x^2 + y^2 - 2ax\cos\theta - 2ay\sin\theta + a^2\cos^2\theta = 0.$$

7. If $k\cos\alpha° = -1$ and $k\sin\alpha° = 1$ and $k > 0$, find k and α.

8. State the maximum and minimum values of the function f defined by $f(x) = 2(\cos x + 1)^2 - 3$.

9. $f: x \to 2x + 3$ and $g: x \to \cos x$ are functions on the set R.
If $h(x) = g(f(x))$, find h', the derivative of h with respect to x.

10. Write down an expression for the shaded area.

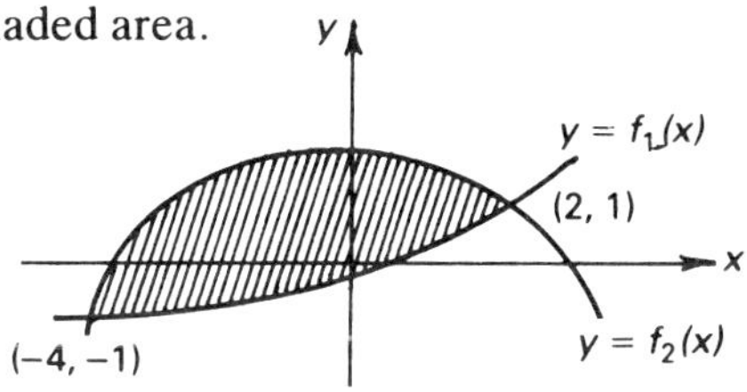

SET 59

1. Evaluate
(i) $\lim_{h \to 0} (2x^2 + 2h)$ (ii) $\lim_{h \to 0} (x + 3)(x - h)$.

2. Find the solution set of the equation $\cos 2x° = \dfrac{\sqrt{3}}{2}$, $0 \leqslant x < 360$ without using your calculator.

3. The sum of the first n terms of a sequence is $3n^2 + 1$. Find the sum of the next n terms.

4. Evaluate $\log_5 125 + \log_8 64$.

5. Sketch the graph of $x^2 + y^2 = 4$. Shade the region which represents $\int_0^2 \sqrt{4 - x^2}\, dx$

6. If $\overrightarrow{PR} \cdot \overrightarrow{PQ} = -1$ what can be said about the size of $R\hat{P}Q$?
7. ABCD, PQRS is a cuboid with AB = 12 cm, AD = 4 cm and AP = 3 cm. Calculate the length of a space diagonal and the angle it makes with base ABCD.
8. Factorise $\cos^4 A - \sin^4 A$ and simplify it.
9. Calculate the gradient of the tangent to the curve $y = 2\cos x$ at the point given by $x = \frac{\pi}{3}$.
10. If $f(x) = (2x^2 - 1)^3$ find $f'(x)$.

SET 60

1. State the maximum value of the function $7 - x^2$.
2. Calculate the distance between
   (i) A(4, –7) and B(11,17) (ii) P(2, –1, 3) and Q(0, 2, –3).
3. The nth term of a sequence is defined by $u_n = n^2 - (u_{n-1})^2$; $u_1 = -1$. Write down the first four terms.
4. Show that the equation $x^3 - 6x + 1 = 0$ has a root between 0 and 1.
5. Find the equation of the sphere centre (2, 1, 3) and radius 5.
6. X is the point (3, 4, 1), Y is (9, 1, –5) and Z is (11, 0, –7). Prove that X, Y and Z are collinear.
7. $f : x \rightarrow \cos x + \sin x$. State the period of f.
8. $f(x)$ is an **even** function if, for all x, $f(-x) = f(x)$. Which of the following are **even**: (i) $\sin x$ (ii) $x \sin x$ (iii) $x + \sin x$?
9. $f(x) = x^2 + x$. Find $f(f'(x))$ where f' is the derivative of f.
10. Given that $g(x) = \sin(1 + \sin x)$ find $g'(x)$.

SET 61

1. Sketch without calculation the graphs of
   (i) $y = x^2 + 1$ (ii) $y = (x + 1)^2$.
2. Write down the equation of the line through P(h, k) parallel to the y-axis.
3. Complete the following implication:
   If $x > 6$, then $x > 3 \Rightarrow$ If $x \leqslant 3$, then . . .
4. State the condition for $x - p$ to be a factor of $f(x)$.
5. The point (h, k) lies inside the circle $x^2 + y^2 = r^2$. State a relationship connecting h, k and r.

6. A is (a_1, a_2, a_3) and B is (b_1, b_2, b_3). Write down the component form of $\underline{a} \cdot \underline{b}$.

7. Simplify
(i) $\sin(90° - A)$ (ii) $\cos(180° + A)$ (iii) $\cos(270° + A)$ (iv) $\sin(180° - A)$.

8. State the most likely function g.

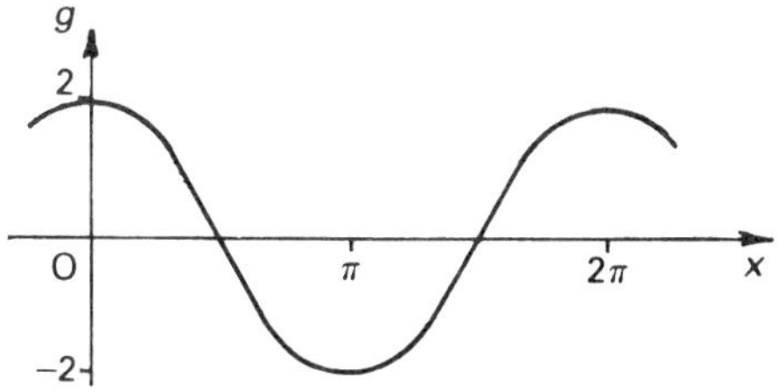

9. Draw a sketch of a function which is **not** differentiable for all x.

10. Sketch the curve $y = \sin 2x$. Shade the area represented by

$$\int_0^{\pi/2} \sin 2x \, dx.$$

SET 62

1. Simplify (i) $(2x)^{-1}$ (ii) $2x^{-1}$

2. Simplify (i) $1 - \cos^2 A$ (ii) $1 - \sin^2 A$.

3. Given that $\log_a b = x$, express b in terms of a and x.

4. A function f defined on the set R is given by $f(x) = \dfrac{ax^2 + bx + c}{(x-k)(x-1)}$.
State any restrictions which must be placed on x.

5. If $\overrightarrow{OB} - \overrightarrow{OA} = \overrightarrow{OC} - \overrightarrow{OD}$, what kind of figure is ABCD?

6. Write down an expression for the distance between $A(x_1, y_1, z_1)$ and $B(x_2, y_2, z_2)$.

7. Write down an expression for:
(i) $\cos^2 A$ in terms of $\cos 2A$ (ii) $\sin^2 A$ in terms of $\cos 2A$.

8. Express $f(x) = \cos x + \sin x$ in the form $k \cos(x - \alpha)$.

9. If $f(x) = x + (k - c)^3$, where k and c are constants, find $f'(x)$.

10. Given that $\int f(x) \, dx = F(x) + c$, evaluate $\int_a^b f(x) \, dx$.

SET 63

1. If x varies as the cube of y and inversely as the square root of z express the constant of variation in terms of x, y and z.

2. Write down the co-ordinates of the image of $P(a, b)$ under reflection in line $y = x$.

3. State the converse of the implication "All dargy splogs are crunky".
   [Hint: Rewrite in "If ... then ..." form]

4. State the condition for the equation $px^2 + qx + r = 0$ to have no real roots.

5. Write down an expression for the length of the vector $\begin{pmatrix} a \\ b \\ c \end{pmatrix}$

6. R is the point (r_1, r_2, r_3) and S is (s_1, s_2, s_3).
   (i) Give the co-ordinates of T, the mid-point of RS.
   (ii) Write down $\underset{\sim}{r} \cdot \underset{\sim}{s}$ in component form.

7. Given that $225° = k\pi$ radians, $k \in R$, find the value of k.

8. (i) State the most likely function f.
   (ii) State the period of function f.

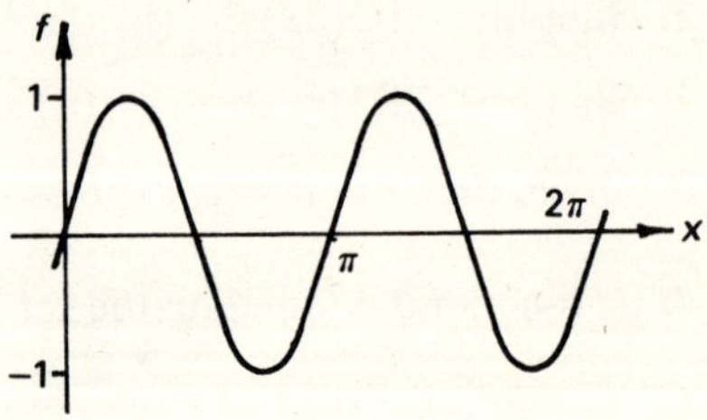

9. Differentiate $\sin^2 5x$.

10. Integrate $\sin^2 5x$.

SET 64

1. y varies directly as x^2 and inversely as z. Write down a formula connecting y, x and z.

2. Give the maximum value of (i) $\sin x$ (ii) $\sin 2x$ (iii) $\sin \frac{x}{2}$.

3. If $p + q = 2$ and $pq = 1$ find the value of $\frac{1}{p} + \frac{1}{q}$.

4. Copy and complete $a^x = b \Leftrightarrow \log$

5. A circle passes through the points P(a, 0), Q(0, b) and the origin. State the co-ordinates of the centre of the circle.

6. If $\overrightarrow{PQ} = k\overrightarrow{QR}$ what can be said about P, Q and R?

7. Simplify
(i) $\sin^2 3A + \cos^2 3A$ (ii) $\sin^2 A - \cos^2 A$ (iii) $2 \cos \theta \sin \theta$.

8. Sketch the graph of the function $\cos 2x$, $0 \leqslant x \leqslant 2\pi$.

9. State $\frac{d}{dx}(\cos x) + \frac{d}{dx}(\sin x)$.

10. If $\int_a^b f(x)\,dx = k$ and $\int_b^c f(x)\,dx = l$, find $\int_a^c f(x)\,dx$.

SET 65

1. Sketch, without calculation, the graph of
(i) $y = x^2$ (ii) $y = x^3$.

2. Write down the equation of the line through P(k, h) perpendicular to the line $y = c$ [k, h, c are constants].

3. Evaluate (i) a^0 (ii) $\log_a a$ (iii) $\log_a 1$.

4. Given the parametric equations $x = t$, $y = t^2 + t$, write down the equation connecting x and y.

5. $\underline{i}$, $\underline{j}$ and $\underline{k}$ are unit orthogonal vectors. Evaluate $\underline{k} \cdot \underline{k}$ and $\underline{k} \cdot \underline{i}$.

6. A is (x_1, y_1) and B is (x_2, y_2). Line AB makes angle θ with OX. Write down an expression for $\tan \theta$.

7. Simplify
(i) $\cos A \cos B - \sin A \sin B$ (ii) $\cos X \sin Y - \sin X \cos Y$.

8. State the maximum value of
(i) $2 \cos x$ (ii) $3 \sin (x - 20°)$ (iii) $\frac{1}{2 - \cos x}$.

9. State an expression for the gradient of the curve $y = f(x)$ at point $(a, f(a))$.

10. The graph of the function f shown has half-turn symmetry about the origin.

Evaluate $\int_{-1}^{1} f(x)\,dx$.

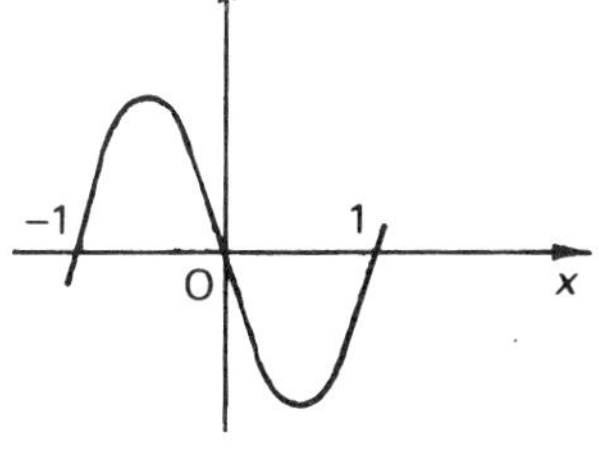

SET 66

1. Write down the 'quadratic formula' for the solution of the equation $hx^2 + kx + l = 0$.

2. A is (x_a, y_a) and B is (x_b, y_b). State the co-ordinates of the mid-point of AB.

3. Copy and complete:
(i) $a^{-n} = \ldots$ (ii) $a^{1/n} = \ldots$ (iii) $\log_a X - \log_a Y = \ldots$ (iv) $\log_b XY = \ldots$

4. A sequence is defined by the recurrence relation $u_{r+1} = 3u_r$. Give a formula for the nth term given $u_1 = 1$.

5. State the equation of the straight line through (h, k) with gradient n.

6. The point $P(\underline{p})$ divides MN in the ratio 2 : 1. Express $\underline{p}$ in terms of $\underline{m}$ and $\underline{n}$.

7. Simplify
(i) $\cos(A + B) - \cos(A - B)$ (ii) $\sin(A - B) - \sin(A + B)$.

8. State the period of

(i) $\tan x$ (ii) $\tan 2x$ (iii) $\sin 3x$ (iv) $\cos\frac{x}{2}$ (v) $\sin\frac{x}{4}$.

9. If $f(x) = \sin x$, state (i) $f'(x)$ (ii) $\int f(x)\,dx$.

10. Define accurately the area enclosed by the curve $y = g(x)$ shown, the x-axis and lines $x = a$ and $x = b$.

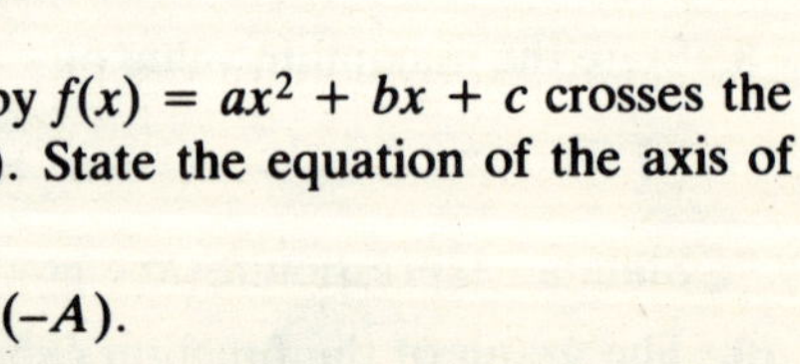

SET 67

1. The graph of the function f defined by $f(x) = ax^2 + bx + c$ crosses the x-axis at points A(–2, 0) and B(4, 0). State the equation of the axis of symmetry of the graph.
2. Simplify (i) $\sin(270° - A)$ (ii) $\cos(-A)$.
3. Express the equation $y = ax^n$ in equivalent logarithmic format.

4. Rationalise the denominators in the expressions (i) $\frac{1}{\sqrt{a}}$ (ii) $\frac{1}{\sqrt{a} - b}$.

5. A radius OP of a circle, centre O, has gradient m. State the gradient of the tangent at P.

6. $\underset{\sim}{u} = \begin{pmatrix} a \\ b \\ c \end{pmatrix}$ and $\underset{\sim}{v} = \begin{pmatrix} d \\ e \\ f \end{pmatrix}$ are two parallel vectors. Write down a relationship connecting the components of $\underset{\sim}{u}$ and $\underset{\sim}{v}$.

7. Sketch the graph of $\sin 2x°$, $0 \leqslant x \leqslant 360$.

8. If $k \cos \alpha = 4$ and $k \sin \alpha = -3$ state the value of k, given that $k > 0$.

9. Describe $f'(x)$ when $f(x)$ is increasing.

10. Evaluate $\int \sin 2x \, dx$.

SET 68

1. State the value for x for which the function $f(x) = \dfrac{1}{x-3}$ is undefined.

2. Describe a regular tetrahedron.

3. Evaluate $\dfrac{\log_5 32}{\log_5 8}$

4. The expression $f(x) = ax^2 + bx + c$ is divided by $x - h$. State an expression for the remainder.

5. The circle $x^2 + y^2 + 2gx + 2fy + c = 0$ passes through the origin. State the value of c.

6. Two vectors $\underset{\sim}{a}$ and $\underset{\sim}{b}$ are at right angles. State the value of $\underset{\sim}{a} \,.\, \underset{\sim}{b}$.

7. Express
   (i) $\cos 2A$ in terms of $\cos A$ (ii) $\cos 4A$ in terms of $\cos 2A$
   (iii) $\cos A$ in terms of $\sin \dfrac{A}{2}$.

8. Write down the maximum value of
   (i) $\cos x$ (ii) $2 \cos x$ (iii) $-2 \cos x$.

9. State an expression for the rate of change of the function $y = f(x)$ when $x = b$.

10. Evaluate $\int_0^2 k dx$, $k \in R$.

SET 69

1. The graph of the function $f : x \rightarrow a^x$ cuts the y-axis at the point A. State the co-ordinates of A.

2. Define a radian.

3. Simplify $\log_a \text{X} + \log_a \text{Y} + \log_a \text{Z}$.

4. State a formula for the $(n + 1)$th term of the sequence $a, ar, ar^2, \ldots$
5. The centre of a circle is (u, v) and the radius is r. Write down the equation.
6. State whether the following are vectors or scalars:
   (i) $\underline{b} \cdot \underline{c} + \underline{c} \cdot \underline{a}$ (ii) $(\underline{b} \cdot \underline{c})\underline{a} + (\underline{c} \cdot \underline{a})\underline{b}$ (iii) $(\underline{b} \cdot \underline{c})(\underline{c} \cdot \underline{a})\underline{b}$.
7. Simplify
   (i) $\cos^2\frac{x}{3} + \sin^2\frac{x}{3}$ (ii) $\cos^2\frac{x}{2} - \sin^2\frac{x}{2}$ (iii) $2\sin\frac{x}{2}\cos\frac{x}{2}$.
8. State the period of (i) $\cos x$ (ii) $\cos 2x$ (iii) $\cos 3x$.
9. $f(x) = (ax + b)^n$. Write down $f'(x)$.
10. Simplify $\int_a^b f(x)\,dx + \int_b^c f(x)\,dx$.

SET 70

1. If $f(x) = x^3 - 6x^2 + 8x$, write down, from the graph, an expression for $g(x)$.

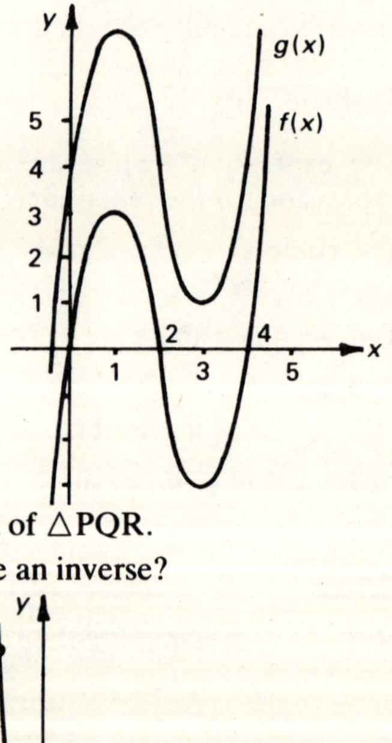

2. State a formula involving $\hat{P}$ for the area of $\triangle$PQR.
3. For a function f, $f(a) = f(b)$. Can f have an inverse?
4. Find an expression for $f(x)$.

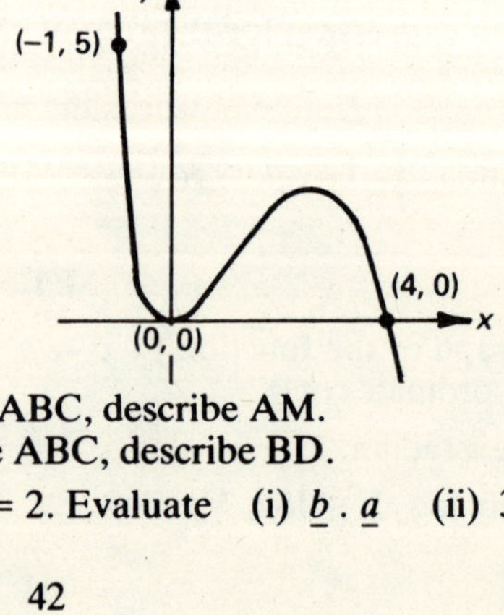

5. (i) AM is a median of triangle ABC, describe AM.
   (ii) BD is an altitude of triangle ABC, describe BD.
6. $\underline{a}$ and $\underline{b}$ are vectors where $\underline{a} \cdot \underline{b} = 2$. Evaluate (i) $\underline{b} \cdot \underline{a}$ (ii) $3\underline{a} \cdot \underline{b}$.

7. The graph is of the function f defined by $f(x) = k + \sin x$. State the value of k.

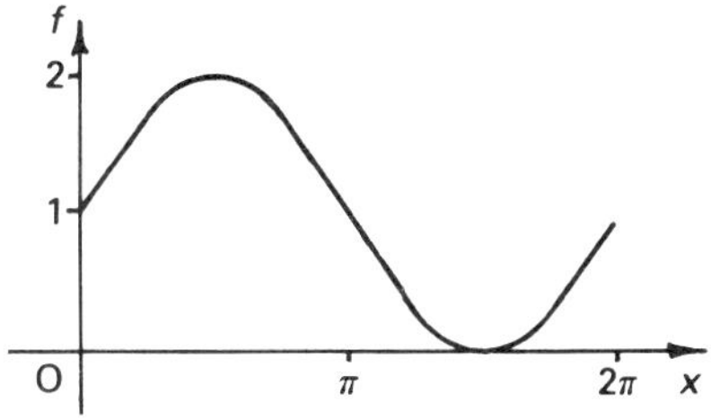

8. Write down the **exact** value of (i) $\cos 60°$ (ii) $\tan 135°$, without using your calculator.

9. State the condition for the curve $y = f(x)$ to have a stationary point at $(a, f(a))$.

10. Integrate (i) $2 \sin x \cos x$ (ii) $2 \cos^2 x - 1$.

SET 71

1. Simplify (i) $2x^0$ (ii) $(2x)^0$.

2. A set of lines passes through (0, 4). State an expression for the equations of the lines.

3. The roots of a quadratic equation are real. Describe the discriminant.

4. Copy and complete (i) $\log_a \frac{1}{n} = \ldots$ (ii) $\log_a X^n = \ldots$

5. Write down the co-ordinates of the centre and length of the radius of the circle $x^2 + y^2 + 2hx + 2ky + l = 0$.

6. Two vectors $\underline{u}$ and $\underline{v}$ are such that $\underline{u} \cdot \underline{v} = 0$. Give three possible interpretations of this result.

7. Is $\frac{1}{3}\underline{i} - \frac{2}{3}\underline{j} + \frac{2}{3}\underline{k}$ a unit vector? Explain.

8. a and b are positive whole numbers and $a \sin x = b$. If the solution set of this equation is $\emptyset$, state a relation connecting a and b.

9. For a function f, state in words the meaning of $f'(1)$.

10. Integrate (i) $(ax + b)^n$ (ii) $\sin (ax + b)$.

SET 72

1. Given that $0 \leqslant x \leqslant 2\pi$ state the range of $3x$.

2. Express $\cos A$ in terms of a, b and c.

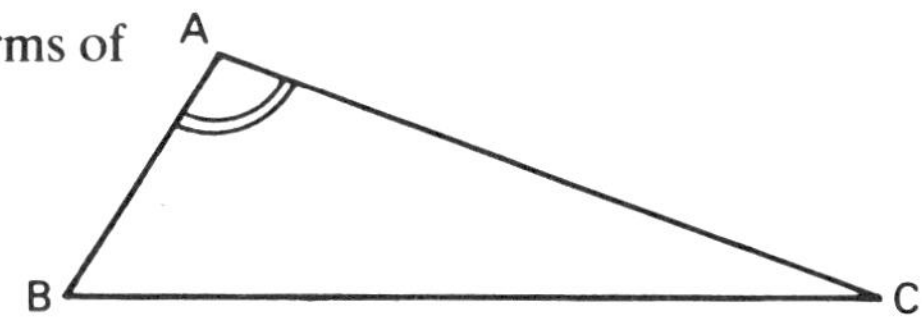

3. Write down the converse of the statement "All bus drivers are men".

4. Express P in terms of a, Q and k, given that $\log P = a \log Q + \log k$.

5. Two lines have gradients m and n. If $mn = -1$, what may be deduced about the lines?

6. Express $\cos\theta$ in terms of $\underline{a}$ and $\underline{b}$.

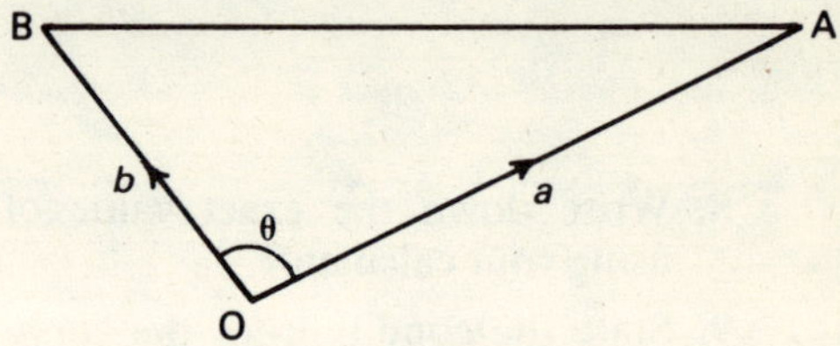

7. Write down the period of (i) $\sin x$ (ii) $2 + \sin x$ (iii) $2\sin x$.

8. Solve the equation $\cos^2 x° = \frac{1}{4}$, $0 \leqslant x \leqslant 360$ without using your calculator.

9. State the condition for $(a, f(a))$ to be a stationary point on the curve $y = f(x)$.

10. Integrate with respect to x (i) $\cos \pi$ (ii) $\cos(\pi - x)$.

SET 73

1.

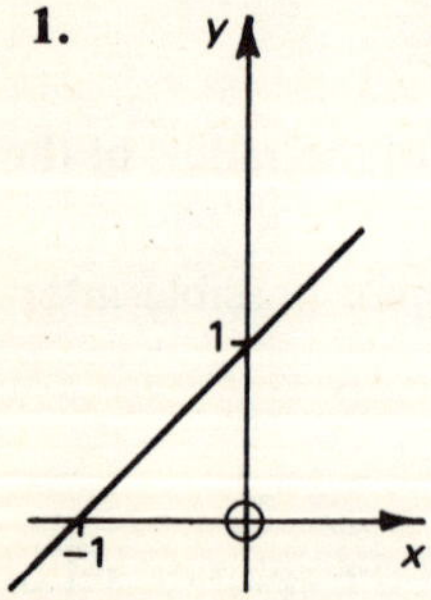

2.

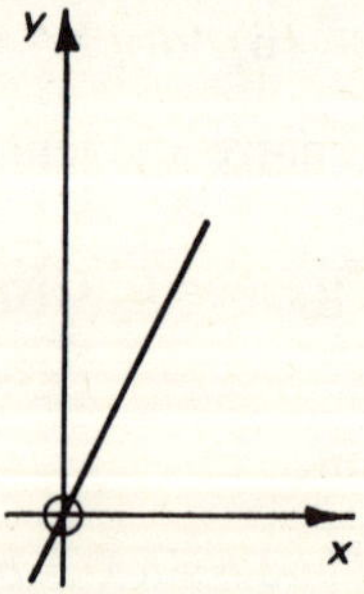

3.

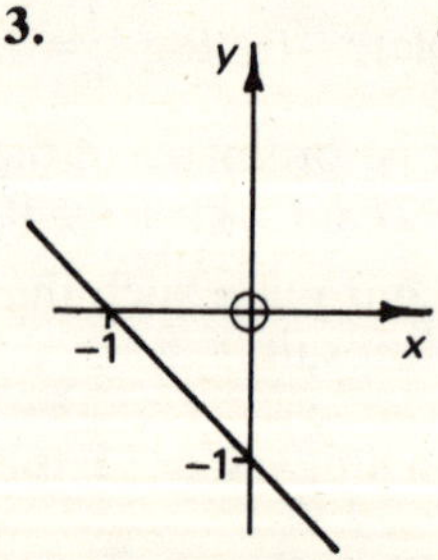

4.

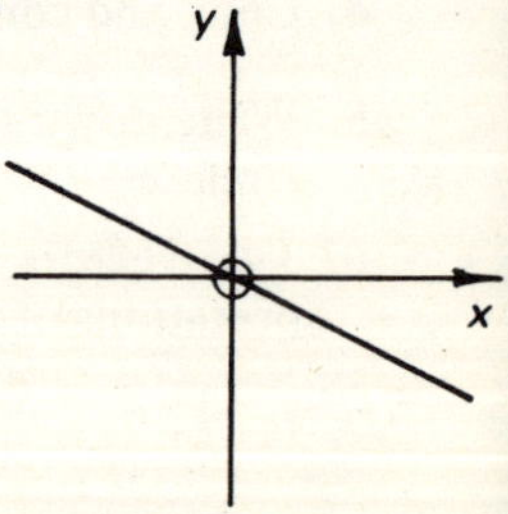

5.

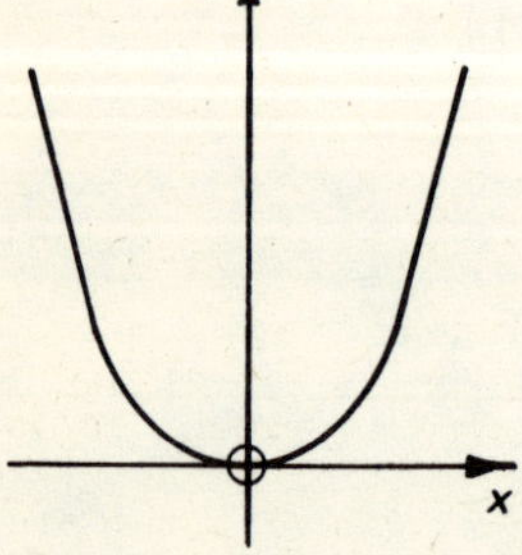

6.

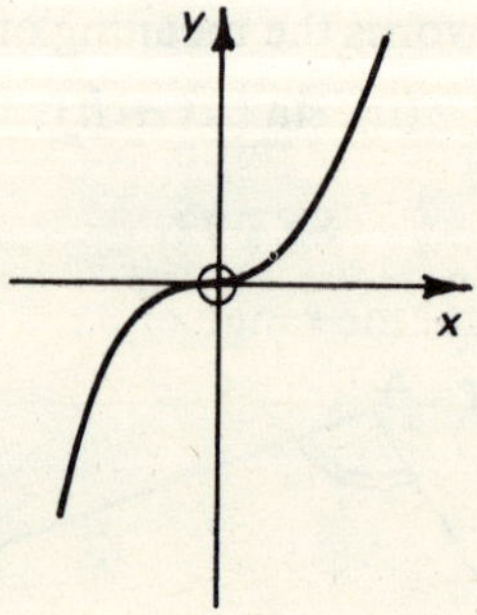

7.

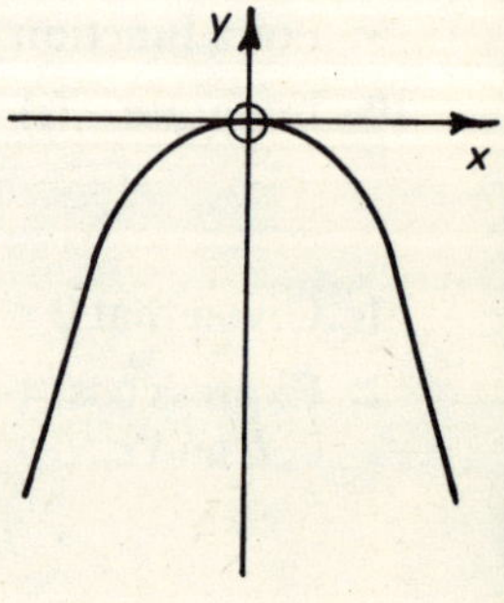

8.

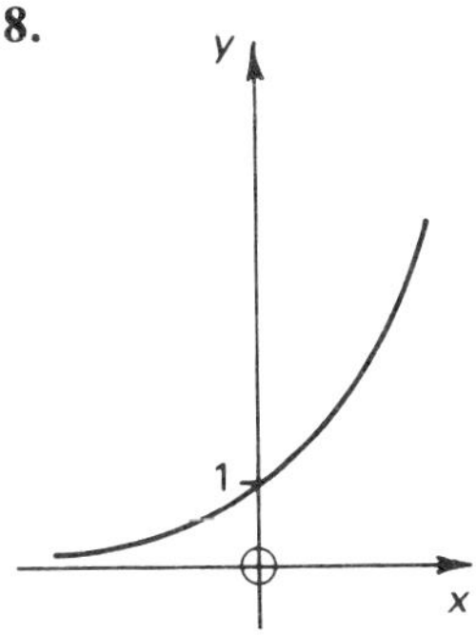

9

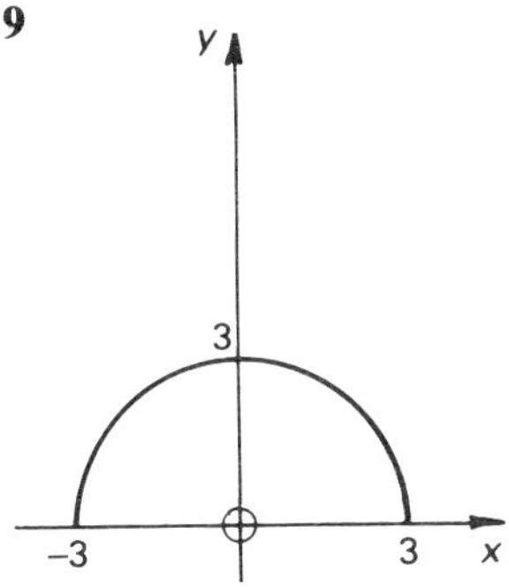

10.

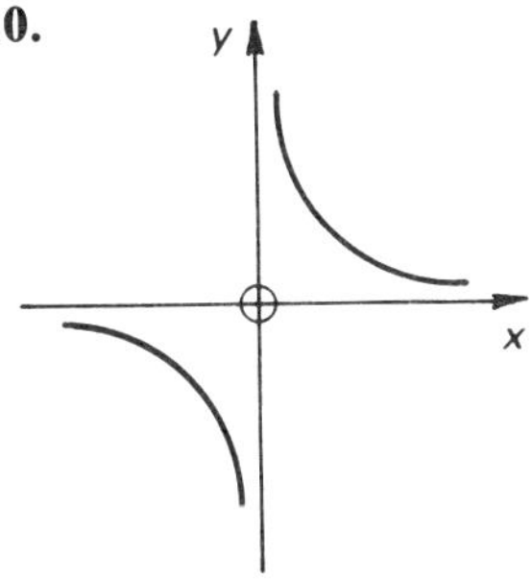

For each of the sketches given above match the graph with the **most likely** equation given in the following list:

(a) $y = a^x, a > 1$
(b) $y = 2x$
(c) $y = -ax^2, a > 0$
(d) $x + y = 1$
(e) $xy = k, k > 0$
(f) $x = y - 1$
(g) $y = x^3$
(h) $2y = -x$
(i) $y = (x + 3)(x - 3)$
(j) $x + y + 1 = 0$
(k) $y = \sqrt{9 - x^2}$
(l) $y = x^2$

SET 74

State whether each of the following is **true** or **false**:

1. $(a + b)^{-n} = a^{-n} + b^{-n}$
2. $2^3 \times 3^4 = 6^7$
3. $\log(X + Y) = \log X + \log Y$
4. $\int \sin^2 x\, dx = \frac{\sin^3 x}{3} + c$
5. $\frac{\log_a M}{\log_a N} = \log_a \frac{M}{N}$.
6. $\sin 4x = 4 \sin x \cos x$.
7. If $f(x) = 3x^2 + 7$ then $f'(x) = 6x + 7$.
8. $\int_a^b f(x)\, dx = -\int_b^a f(x)\, dx$
9. $\int_a^b g(x)\, dx + \int_b^c g(x)\, dx = \int_a^c g(x)\, dx$
10. $(1, -1)$ is a turning point on the curve $y = x^2 - 2x$.

SET 75

This graph of the quadratic function f refers to questions 1 to 6.

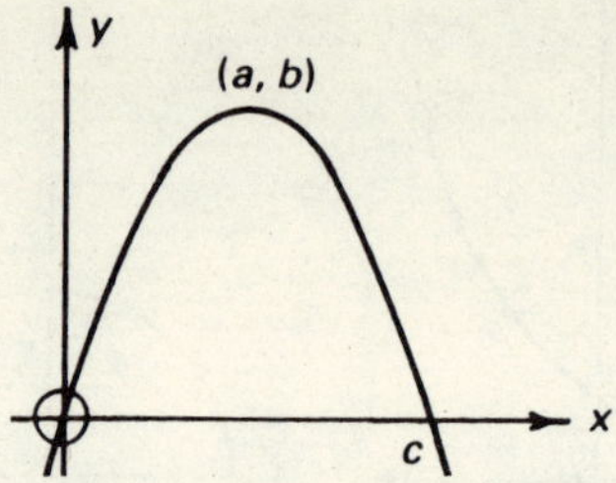

Given the graph of f above, sketch the graphs of the following related functions, annotating each as fully as the limited given information will allow:

1. $y = -f(x)$

2. $y = f(-x)$

3. $y = -f(-x)$

4. $y = f(x) + 2$

5. $y = f(x + 2)$

6. $y = f'(x)$

For each function described by its graph in questions 7 to 10, draw a sketch of the derived function, annotating each as fully as the given information will allow:

7.

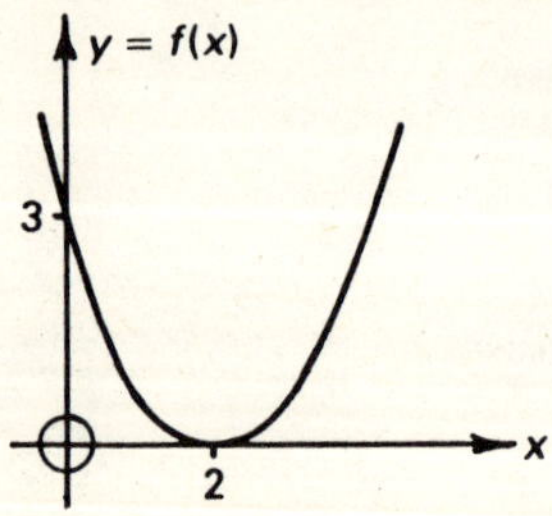

8.

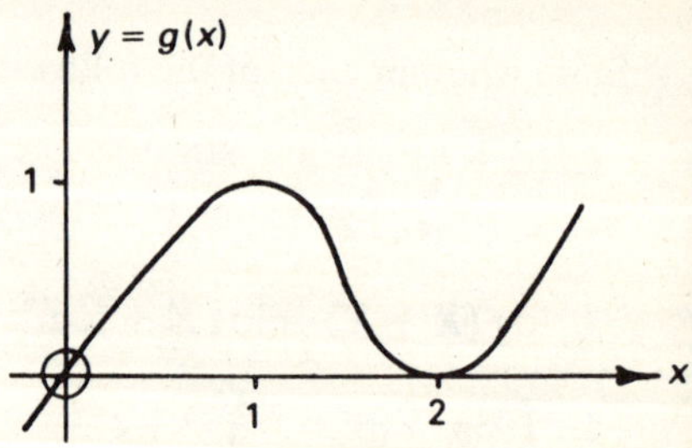

9.

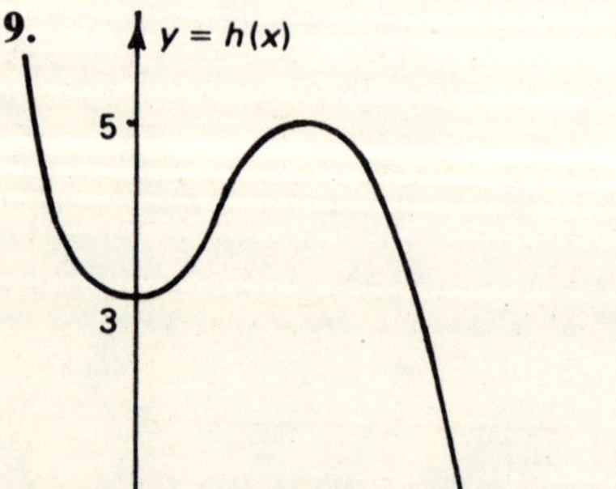

10.

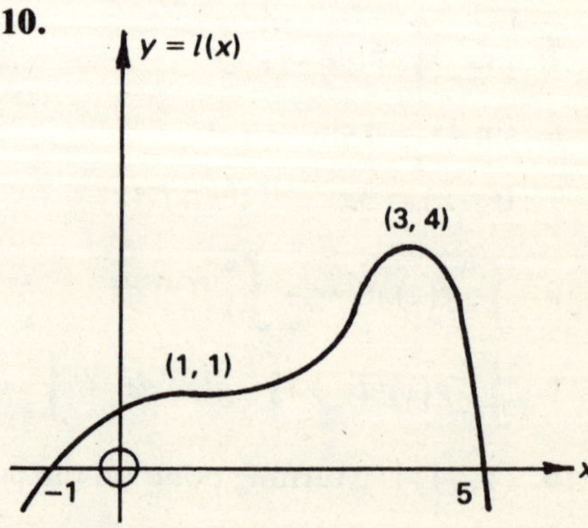

SET 76

Copy the graphs of the functions f, g, h and l in questions 1 to 4 and on each diagram add the graph of the appropriate inverse function f^{-1}, g^{-1}, h^{-1} or l^{-1} respectively.

1.

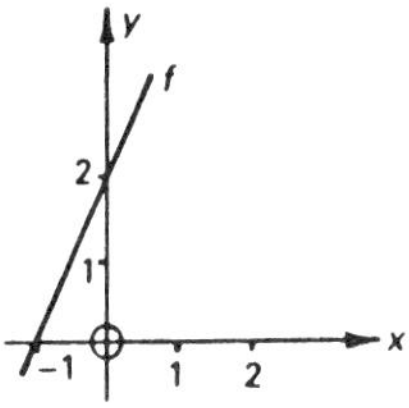

2.

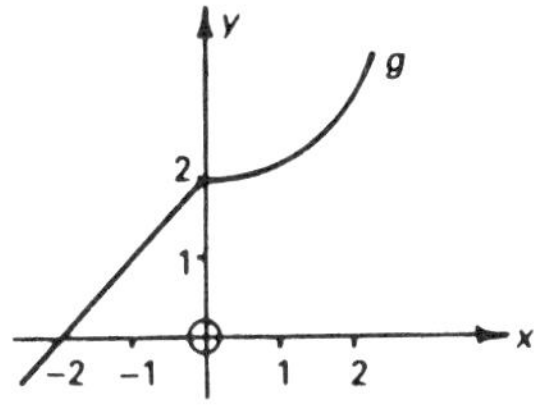

3.

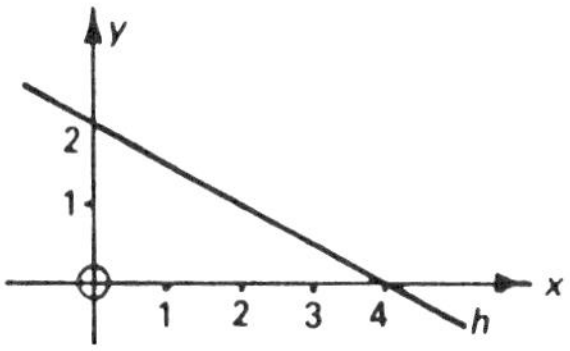

4.

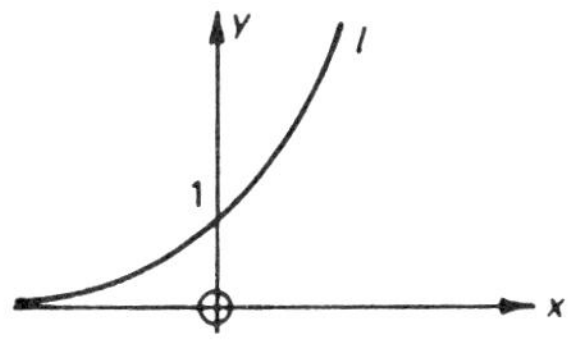

The graphs in questions 6 to 10 are related to the graph of f shown here. Identify these graphs. [Hint: look back at Set 75.]

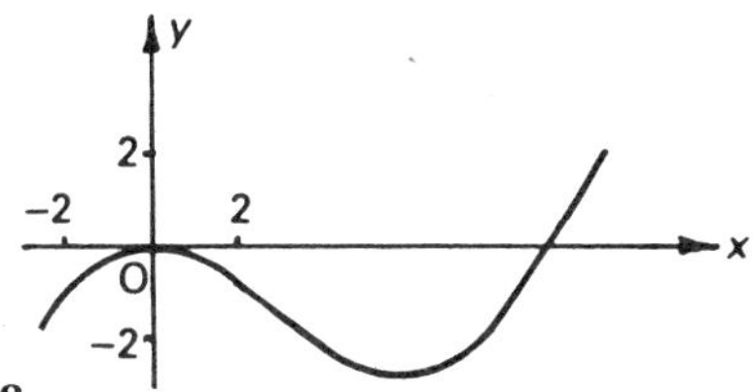

5.

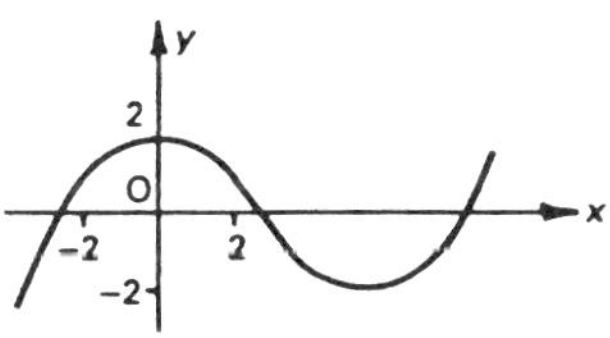

6.

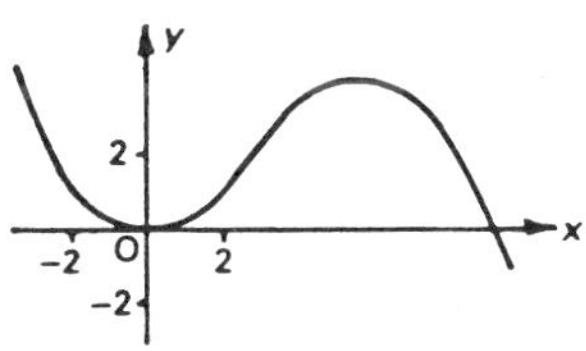

7.

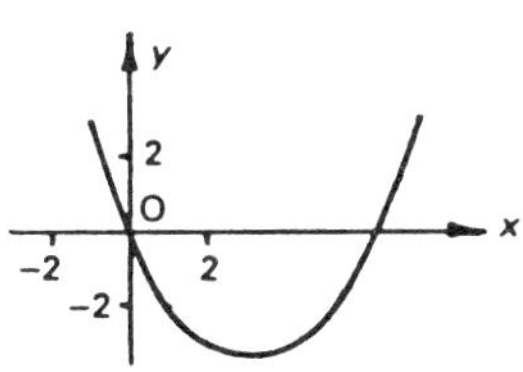

8.

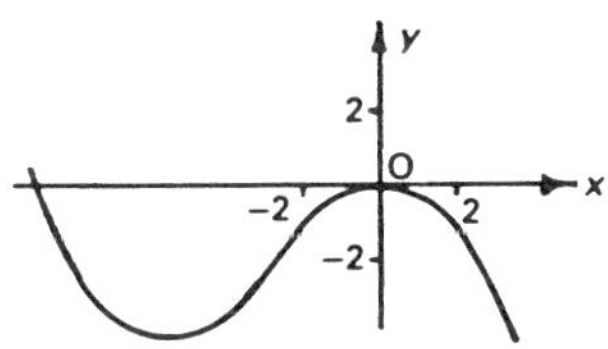

9.

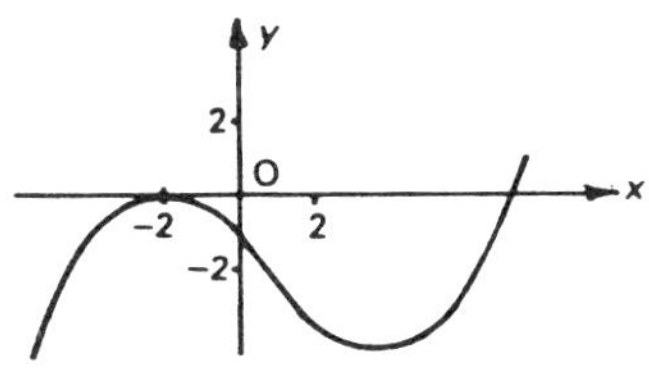

10.

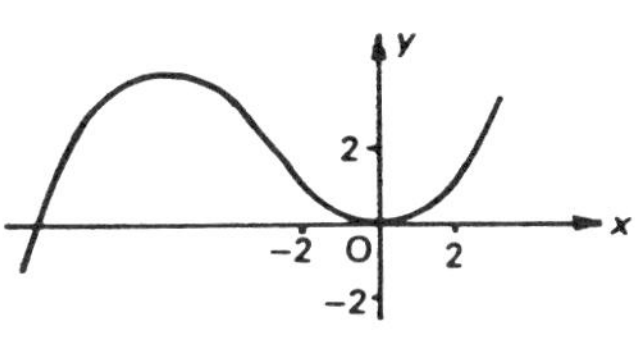

SET 77

1. Write down the highest common factor of 99 and 9999.

2. Write down the largest multiple of 7 less than one million.

3. Write down the units digit in the answer to
(a) 2^{100} *(b)* $(9864)^3 + (8963)^4$

4. If x and y are positive integers such that $y^2 = 12x$, state the least possible value of x.

5. The terms of a sequence satisfy the recurrence relation $u_{n+1} = 20 + \frac{3}{4}u_n$.

Evaluate $\lim_{n \to \infty} u_n$.

6. A cup of tea costs 4p and a cup of coffee 5p in a subsidised canteen. A person spends exactly k pence on only cups of tea and coffee. For what value of k between 8 and 28 inclusive is this impossible?

7. Think of Rubik's cube as a cube made up of 27 congruent smaller cubes. If the volume of a smaller cube is 8 cm^3, find the area of one face of Rubik's cube.

8. The vertex of each smaller square is at the mid-point of a side of a larger square.
What fraction of the figure is shaded?

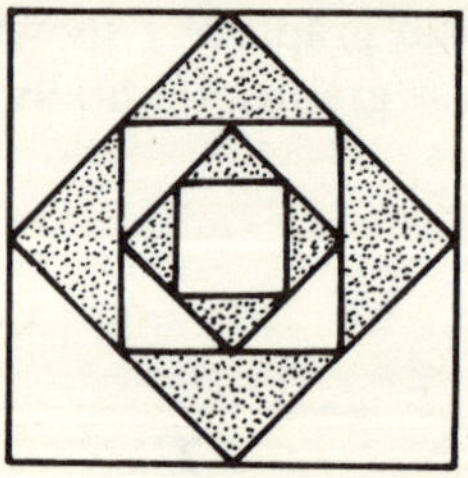

9. If the sides of the square measure a units and all the radii are equal, prove that the shaded area is $\frac{a^2}{4}(4-\pi)$ units^2.

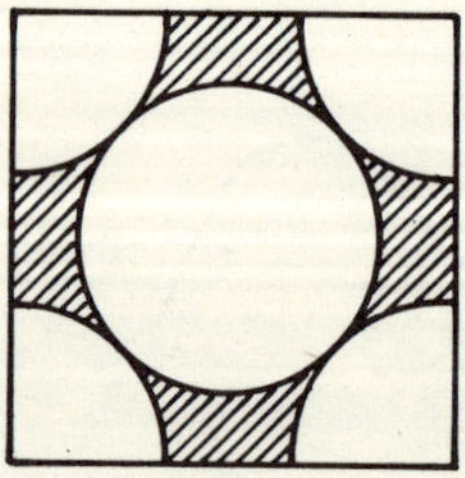

10. Prove that *abba*, any four digit palindromic number (one which reads the same backwards as forwards) is divisible by 11.

ANSWERS

SET 1

1. $x^2 - 11x + 5$ **2.** $4\sqrt{2}$ **3.** 0 **4.** $-2{\cdot}32$ or $4{\cdot}32$
5. $m_{AB} = 2 = m_{BC}$ **6.** $(2, -3)$; 4 **7.** $\cos x$ **8.** 30, 210
9. 20 **10.** $x^4 - x^2 + x + c$

SET 2

1. $x^3 + 4x^2 + x - 6$ **2.** $(9, 7)$ **3.** 16 **4.** $\{-2, -1, 0, 1, 2\}$
5. $(8, 6)$ is on all 3 **6.** $(4, 9)$ **7.** $\sin^2 A$ **8.** $\sqrt{3}$
9. -6 **10.** $8x(x^2 + 4)^3$

SET 3

1. $\dfrac{4 - 2\sqrt{3}}{\sqrt{3}} = \dfrac{4\sqrt{3} - \sqrt{6}}{3}$
2. $AB = BC$, $m_{AB} \cdot m_{BC} = -1$ or $AC^2 = AB^2 + BC^2$
3. 4; 31·5 **4.** $2x^2 + (2c - 9)x + c^2 = 0$ **5.** $5x - 3y - 9 = 0$
6. $x^2 + y^2 - 6x - 2y - 6 = 0$ **7.** $\dfrac{\sqrt{3} - 1}{2\sqrt{2}} = \dfrac{\sqrt{6} - \sqrt{2}}{4}$
8. (i) $\left\{\dfrac{\pi}{6}, \dfrac{5\pi}{6}, \dfrac{7\pi}{6}, \dfrac{11\pi}{6}\right\}$ (ii) $\emptyset$ **9.** $4x^3 + 12x$ or $4x(x^2 + 3)$
10. $\frac{1}{3}x^3 + 2x^2 + 4x + c$ or $\frac{1}{3}(x + 2)^3 + k$

SET 4

1. *(a)* $r = \dfrac{v - u}{t}$ *(b)* $F = \dfrac{9}{5}C + 32$ *(c)* $g = \dfrac{u^2 - v^2}{2\,s}$ **2.** $a = \pm 3$
3. -6 **4.** 0 or 5 **5.** (i) -4 (ii) $\dfrac{-2}{3}$ **6.** $\begin{pmatrix} 1 \\ -1 \end{pmatrix}$ **7.** $2\sin\theta$
8. $\dfrac{3\pi}{8}$ **9.** 2 **10.** $\dfrac{x}{\sqrt{x^2 + 1}}$

SET 5

1. $x > \dfrac{4}{5}$ **2.** (i) $(a, -b)$ (ii) $(-a, -b)$ (iii) $(b, -a)$ (iv) (b, a)
3. $\dfrac{2}{5}$ **4.** $\{x: -3 < x < 8\}$ **5.** -8 **6.** $\dfrac{4}{3}$ **7.** $2\sin A\cos B$
8. 35·3° **9.** $\frac{1}{2}u^{-1/2} - \frac{9}{2}u^{1/2}$ **10.** $f'(x) = 2x$

$$\int f(x)\,dx = \frac{1}{3}x^3 + c$$

SET 6

1. (i) $2x(x-4y)$ (ii) $(x-6)(x+1)$ (iii) $x(x+1)(x-2)$ **2.** $\dfrac{1}{\sqrt{2}}$

3. $\dfrac{7}{8}$ **4.** (i) 5 (ii) 1 **5.** 53·1° **6.** collinear **7.** 5, –3

8. 1·5 **9.** –1 **10.** $\dfrac{9}{2}x^2(x^3+5)^{1/2}$

SET 7

1. $\left.\begin{array}{l} x = 2 \\ y = -3 \end{array}\right\}$ **2.** $\dfrac{8}{9}$ **3.** 6 **4.** 35 **5.** $\dfrac{-1}{5}$ **6.** (4, –1)

7. 25 **8.** 0·75 units^2 **9.** $2x+8$ or $2(x+4)$ **10.** $y = x^2 + x + 1$

SET 8

1. $\dfrac{x-3}{2}$

2.

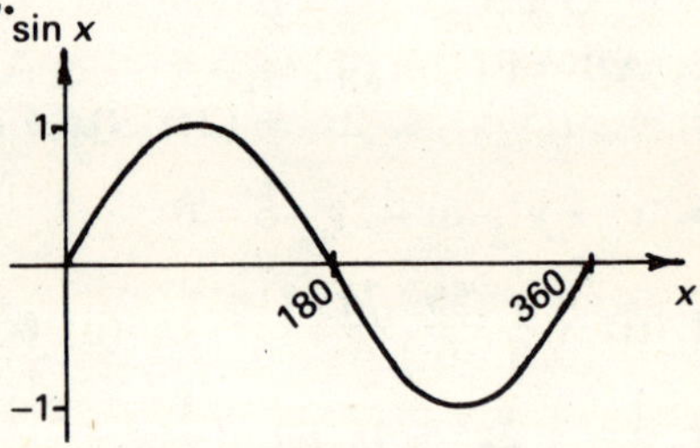

3. $a = 6, p = 3$

4. (i) $\sqrt{3}$ (ii) 4

5. –7 **6.** $\sqrt{61} = 7{\cdot}8$

7. 10 **8.** 1·12

9. 5 **10.** $\dfrac{1}{(1-x)^2}$

SET 9

1. $-7\frac{1}{2}$ **2.** (i) $\dfrac{-3}{5}$ (ii) $\dfrac{-4}{3}$ **3.** 81 **4.** $\left\{x: x \leqslant \dfrac{-2}{3}\right\} \cup \left\{x: x \geqslant 2\frac{1}{2}\right\}$

5. 0 **6.** outside **7.** $\sin(A - C)$ **8.** 41·4° **9.** 39

10. (i) $-2x$ (ii) $c + 9x - \dfrac{1}{3}x^3$

SET 10

1. –5 **2.** 78·5° **3.** 4, –2 **4.** (i) P, (ii) Q, (iii) $\emptyset$

5. $2x - y - 6 = 0$ **6.** $3a = 2b$ **7.** 27° **8.** $22\frac{1}{2}, 67\frac{1}{2}, 112\frac{1}{2}, 157\frac{1}{2}$

9. (0, 0) Pt. of rising inflection **10.** $4x(3x^2-5)^{-1/3}$

SET 11

1. 18 **2.** 36 units^2 **3.** (i) 35, (ii) 39 366 **4.** $b^2 = a^2 \Rightarrow b = \pm a$

5. $3x + 2y = 4$ **6.** $x^2 + y^2 - 8y + 7 = 0$

7. (i) cos 65°, (ii) $\sin 135° = \dfrac{1}{\sqrt{2}}$, (iii) $-\cos 70° = \cos 110°$

8. 45° **9.** $12x - 45x^2$ or $3x(4-15x)$ **10.** $c - \dfrac{1}{x}$

SET 12

1. $2\sqrt{3}$ **2.** $(1, 3)$ **3.** $27, 4x^2 + 6x + 16$ **4.** 27 **5.** $x + y = 2$

6. (i) $\underline{b} - \underline{a}$, (ii) $\underline{m} = \frac{1}{2}(\underline{a} + \underline{b})$ **7.** 0 **8.** $\frac{-2}{\sqrt{6}} = -\frac{\sqrt{6}}{3} = -0{\cdot}816$

9. $x > 2$ **10.** $\frac{-2x}{(1 + x^2)^2}$

SET 13

1. $-12x$ **2.** $7{\cdot}66\ \text{cm}^2$ **3.** $x^2 + 8x + 15$ **4.** $\{x: x < 1\} \cup \{x: x > 2\}$

5. $m_{AB} \cdot m_{AC} = -1$ **6.** $(2, -1); 4$ **7.** $\frac{1}{3}$ **8.** $21{\cdot}8°$

9. (i) -2, (ii) -2, (iii) -2 **10.** $3x^3 - 3x^2 + x + c$ or $\frac{1}{9}(3x - 1)^3 + k$

SET 14

1. (i) $\frac{1}{4}$, (ii) $\frac{2}{a^3}$ **2.** 16 or -2 **3.** -2 **4.** $x^3 - \frac{1}{x^3}$ **5.** $\frac{1}{2}$

6 $m = 6, n = -1$ **7.** (i) $\sin x$, (ii) 1 **8.** Max = 6, min = 4

9. 0 **10.** $\frac{x}{\sqrt{(x + 1)(x - 1)}}$

SET 15

1. $x + 3$ **2.** $2\pi, \pi$ **3.** (i) 91, (ii) 9 **4.** 8 **5.** $m_{AB} = \frac{2}{3} = m_{AC}$

6. Concentric circle with radius 3 **7.** (i) $\cos 2x$, (ii) 1

8. (i) 45°, (ii) $35{\cdot}3°$ **9.** $\frac{-1}{2u^{3/2}}$ **10.** $\frac{2}{5}$

SET 16

1. $\sqrt{120} = 10{\cdot}95$ **2.** (i) $y = 3x + c$ (ii) $y = mx + 5$ **3.** $(x + 3)^2 - 1$

4. (i) ± 2; (ii) -3 **5.** $\frac{-7}{4}$ **6.** $\begin{pmatrix} \frac{1}{2} \\ -4 \\ -1\frac{1}{2} \end{pmatrix}$

7. *(a)* $\cos(-A) = \cos A$; *(b)* $\sin(-A) = -\sin A$; *(c)* $\tan(-A) = -\tan A$; *(d)* $\cos(270° + A) = \sin A$

8. $2\cos^2 A - 1$ **9.** $(1, -1)$ min. t. pt. **10.** $\frac{-2x}{(x^2 - 1)^2}$

SET 17

1. 2 **2.** $4{\cdot}5$ cm **3.** $4x^2 + 4x + 1$ **4.** 1 **5.** $(0, 5)$

6. $\begin{pmatrix} -2 \\ 1 \end{pmatrix}$ **7.** 30, 90 **8.** 45° **9.** 6 **10.** 0

SET 18

1. $\{-4, 9\}$ **2.** $1; -1$ **3.** $(x-1)(x-2)(x+3)$ **4.** $(\pm 2, 2)$
5. $x - y + 2 = 0$ **6.** parallelogram
7. -1 **8.** -1

9. $f(x)$ never decreases **10.** $\dfrac{-1}{2(x+1)^{3/2}}$

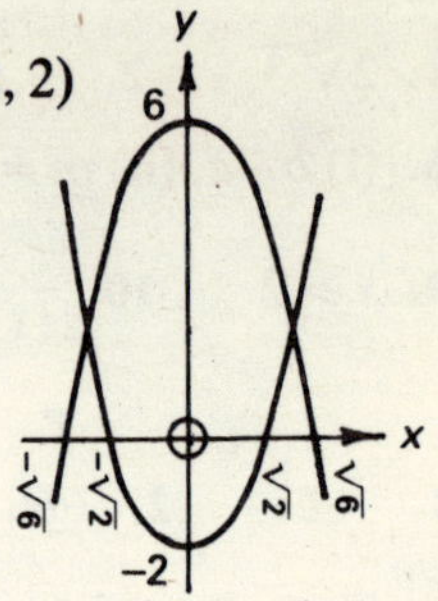

SET 19

1. $4\frac{1}{2}$ **2.** $-\frac{1}{2}$ **3.** $10n$ **4.** ± 1

5.

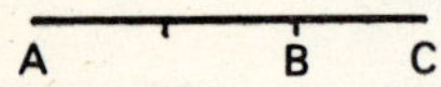

6. ± 3 **7.** $\dfrac{24}{25}$

8. $45°$ **9.** (i) $\dfrac{1}{2\sqrt{u}}$; (ii) $-\dfrac{2}{u^3}$ **10.** $s = t^3 - t - 3$

SET 20

1. $x + 2 + \dfrac{1}{x}$ **2.** $(7, 1)$ **3.** -27 **4.** $\{x : 4 \leqslant x \leqslant 5, x \in R\}$

5. $3x - 2y = 0$ **6.** 6 **7.** $\cos \pi, \cos \dfrac{2\pi}{3}, \sin \dfrac{\pi}{6}, \sin \dfrac{\pi}{2}, \tan \dfrac{\pi}{3}$

8. 0 **9.** $45°$ **10.** $2 \cos 2x$

SET 21

1. 2 **2.** $\{131{\cdot}8, 228{\cdot}2\}$ **3.** $x^2 + 6x + 8; x^2 + 2$
4. $2x^2 + (2p-3)x + (p^2 + 3q) = 0$ **5.** $p = 2$ **6.** $(4, 2); \sqrt{29}$

7. (i) 0; (ii) 1 **8.** $55{\cdot}6°$ **9.** $\dfrac{3}{2}x^{1/2}$ **10.** $18\frac{2}{15}$

SET 22

1. -1 **2.** $\frac{1}{2}$ **3.** $(x+1)(x-2)(x+5)$ **4.** $x = -1\frac{1}{2}$

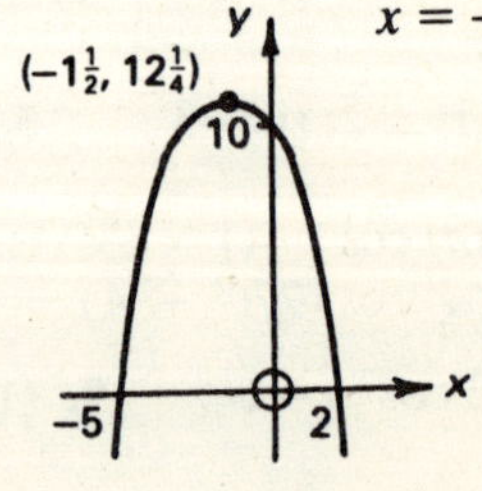

5. 4 units^2
6. (i) $(6, 9)$;
(ii) Both $= 13$ or they are diagonals of a parallelogram

7. $\dfrac{\sqrt{3}+1}{2\sqrt{2}} = \dfrac{\sqrt{6}+\sqrt{2}}{4}$ **8.** 29

9. $x < 1$ **10.** $-5 \sin (5x + 4)$

SET 23

1. (i) ± 2; (ii) 2, 3 **2.** $3x - y + 5 = 0$ **3.** $2, -\dfrac{1}{2}, -\dfrac{1}{6}$ **4.** $\{p, q, r, t\}$

5. $76°$ **6.** $(13, 0)$ **7.** 1 **8.** $210°$ **9.** -12 **10.** $4x - x^2 + c$

SET 24

1. 4 : 1 **2.** $\sin^2 A$ **3.** $k = -1\frac{1}{2}$, $m = 2\frac{1}{4}$ **4.** $x \geqslant -2$ **5.** $2\frac{2}{3}$

6. $\frac{\sqrt{17}}{9}$ **7.** 155 **8.** 45° **9** $f(a)$ and $f(b)$ **10.** $\frac{1}{2(1-x)^{3/2}}$

SET 25

1. $x + 1$ **2.** 6 **3.** –2 **4.** 0, 5 **5.** $4x - 3y - 2 = 0$ **6.** 24 cm

7. $\cos A$ **8.** (i) 90°; (ii) 45° **9.** $4 - \frac{3}{x^2}$ **10.** $y = x$

SET 26

1. 5 **2.** $m = -\frac{5}{4}$; 5 **3.** –1 **4.** {1, 3, 7, 9} **5.** $-\frac{1}{2}$ **6.** 6

7. 130 **8.** (0, 1) (2π, 1) etc. **9.** (i) $2(x - 1)$; (ii) $2x$

10. $2x \cos(x^2 + 2)$

SET 27

1. $f = \frac{uv}{v + u}$ **2.** $b^2 + c^2 - a^2 : a^2 + c^2 - b^2$ **3.** 0, 1, 1, 0, –1, –1

4. $-4 < x < 2$ **5.** –6, 10 **6.** (1, 2) **7.** $-\frac{3}{5}$ **8.** 54·5°

9. $\frac{1}{2x^{1/2}} - \frac{1}{2x^{3/2}}$ **10.** $y = 2x - 4$

SET 28

1. $1 + x$ **2.** A(0, 10), B(4, 0), C(8, –10) **3.** $(x - 2)(x + 2)(x - 3)$

4. $\log_2 y = x$ **5.** $x - y - 5 = 0$ **6.** $\underline{p} + \underline{q} - \underline{r}$ **7.** $k = 5$; $x = 53{\cdot}1$

8. (i) 90°; (ii) 1440° **9.** $f'(a)$ **10.** $-2\sin(2x - 1)$

SET 29

1. (–2, 0), (3, 0) **2.** $(6 \sin y - 5)(\sin y + 1)$ **3.** 0 **4.** 16

5. (–1, 3) **6.** $AB = r_1 + r_2$ **7.** (i) 1; (ii) 1 **8.** 63·4°

9. $1 - \frac{4}{x^2}$ **10.** $5\frac{1}{3}$

SET 30

1. $2^{(3^2)}$ **2.** 11 **3.** 17 **4.** $y = a^p$ **5.** (i) $y = x$; (ii) $y = 2x$

6. –5 **7.** $\sqrt{13}\cos(x - 56{\cdot}3)°$ **8.** {166, 226} **9.** $f'(a)$

10. $\frac{-1}{2(x-2)^2}$

SET 31

1. $\dfrac{-h(2x+h)}{x^2(x+h)^2}$ **2.** (i) {30, 330}; (ii) {120, 300} **3.** 4 **4.** 8

5. $(p, 0)$ **6.** $-\frac{2}{3}$ **7.** $\frac{\pi}{4}$ **8.** 90°

9. $\frac{3}{2}x^{1/2} + x^{-1/2} + \frac{3}{2}x^{-3/2}$ **10.** 10 units^2

SET 32

1. 1 **2.** (i) $(2-a, b)$; (ii) $(a, -4-b)$; (iii) $(-b, a)$ **3.** $8-(x+1)^2$

4. $x = 32$ **5.** (i) $y = 4$; (ii) $x = 3$; (iii) $4x - 3y = 0$ **6.** (i) $\frac{1}{2}$; (ii) $-\frac{1}{2}$

7. $3\sin(x + 48{\cdot}2)°$ **8.** $0, \frac{\pi}{3}, \pi$ **9.** $a = -1, b = 5$

10. $4\cos x(1 + \sin x)^3$

SET 33

1. 23 **2.** 5 **3.** $\sin^2 x$

4. roots are real and distinct

5. 1

6. $x^2 + y^2 + 2gx + 2fy = 0$

7. (i) $2\cos^2\frac{A}{2} - 1$; (ii) $1 - 2\sin^2\frac{A}{2}$

8. 180, 360

9. $x^{2/3} - 2 + x^{-2/3}$

$\frac{2}{3}x^{-1/3} - \frac{2}{3}x^{-5/3}$

10.

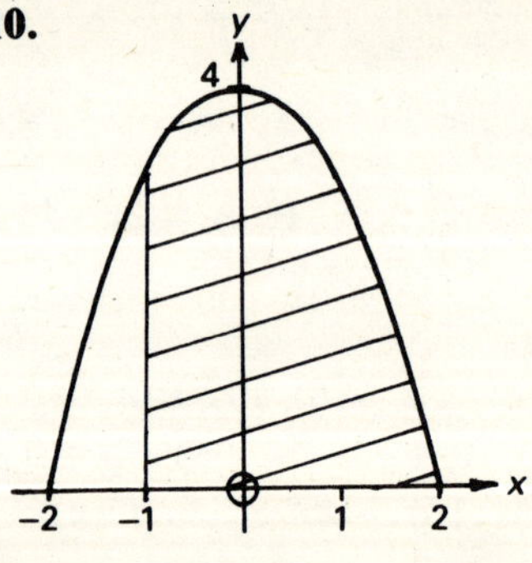

SET 34

1. $(x, y) \in \{(1, 1), (-1, -1)\}$ **2.** (i) $-\sin\theta$; (ii) $\cos\theta$ **3.** 18

4. $S' = \{x : x \leqslant -2, x \in R\}$ **5.** $x - 2y = 6$ **6.** $-4\frac{1}{2}$ **7.** 2

8. $\cos 2A$ **9.** 2 **10.** $-2\cos x \sin x\ (= -\sin 2x)$

SET 35

1. 9 **2.** collinear **3.** 15; 0 **4.** $\{x: x \leqslant -2\} \cup \{x: x \geqslant 6\}$ **5.** $\dfrac{u+t}{2u}$

6. $\left(1, -\frac{7}{4}\right)$ **7.** $\frac{4}{5}$ **8.** 21·8° **9.** $3ax^2 - 2ax$ **10.** 2π

SET 36

1. 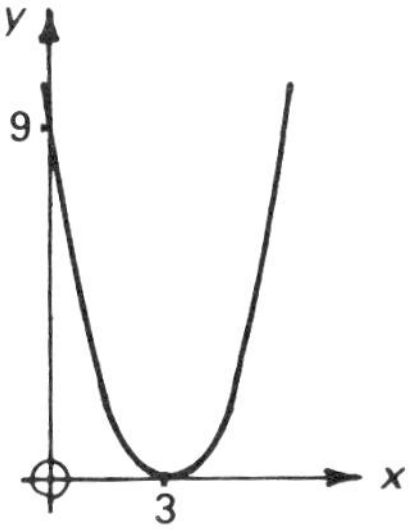

2. $\frac{1}{4}$ 3. $2(x+1)(x-2)(x+3)$
4. $\log_q 3 = n$ 5. -2
6. (i) 18; (ii) -7 7. $-\sqrt{13}$
8. $1 + \cos x$ 9. -10
10. $4x \cos(2x^2 + 1)$

SET 37

1. $10x^2 - 3x$ 2. $(1, -1)$ 3. $x^2 + 6x + 12$
4. $\triangle = k^2 + 20 \geqslant 20 \Rightarrow \triangle$ +ve $\Rightarrow$ roots real
5. (i) $y = 3x + 2$; (ii) $4x + 5y = 10$ 6. $(x-1)^2 + (y+2)^2 = 9$
7. 1 8. 90° 9. 2
10. (i) $18x - 6$ or $6(3x-1)$; (ii) $3x^3 - 3x^2 + x + c$ or $\frac{1}{9}(3x-1)^3 + k$

SET 38

1. 0, 2 2. (i) $(2\sin x - 1)(\sin x + 3)$; (ii) $\sin A(2 - \sin A)$
3. u_5 4. 3 5. $-\frac{11}{2}$ 6. 16 7. 60 8. $\frac{1}{4}$ 9. -10
10. $4(2x+3)(x^2+3x-5)^3$

SET 39

1. 4 2. $2x + y = 3$ 3. 1, -1, -2
4. 85 5. $x - 2y + 9 = 0$
6. $(2, -3)$; 4 7. $-\frac{24}{25}$
8. 32° 9. $V'(r) = \pi r^2$

10. 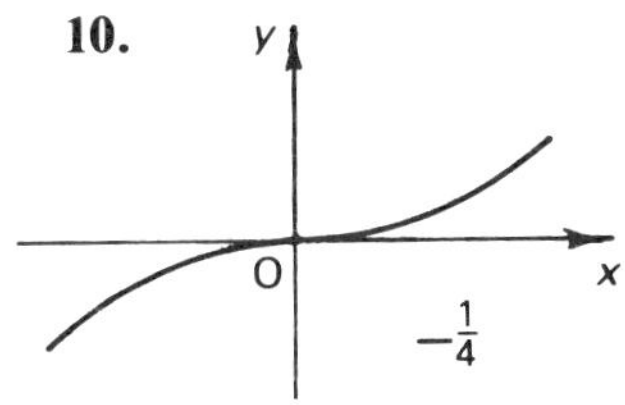

SET 40

1. $\frac{a-1}{a-4}$ 2. 29° 3. $(x^2+1)^2 + 4$; 4 4. 81
5. $m_{AB} = -1 = m_{AC}$ 6. 0 7. 6 8. $\sin 2x$
9. $\frac{dy}{dx} = 3x^2 \not< 0$ 10. $-6x\sin(3x^2 - 5)$

SET 41

1. -1, 4 2. $2x + 3y = 9$ 3. ± 2 4. $-\frac{1}{2}$, 1 5. $x + 2y = 11$
6. $x^2 + y^2 + 2x - 4y - 4 = 0$ 7. $-\frac{3}{5}$ 8. 71·6°
9. (i) $\cos\alpha$; (ii) $2\cos(2\alpha + 1)$; (iii) $-3\cos(2 - 3\alpha)$ 10. 0

SET 42

1. $6xh + 3h^2$ **2.** (i) $6{\cdot}3\ u$; (ii) $18{\cdot}8\ u^2$; (iii) $15{\cdot}6\ u^2$

3. $(x-2)(x+5)(2x-1)$ **4.** 16 **5.** -5 **6.** $\dfrac{\overrightarrow{QP} \cdot \overrightarrow{QR}}{|\overrightarrow{QP}||\overrightarrow{QR}|}$

7. -9 **8.** 1 **9.** 4 **10.** $\frac{3}{2}(2x-5)(x^2-5x-2)^{\frac{1}{2}}$

SET 43

1. $\frac{3}{4}y^{1/6}$ **2.** 130°

3. 2, 11, 101, 1001, 10 001

4. real, distinct

5. $(2\frac{1}{2}, 5)$

6. $x^2 + y^2 - 6x - 6y + 5 = 0$

7. $\sin x$ **8.** $21{\cdot}8°$

9. $\cos x - \sin x$

10.

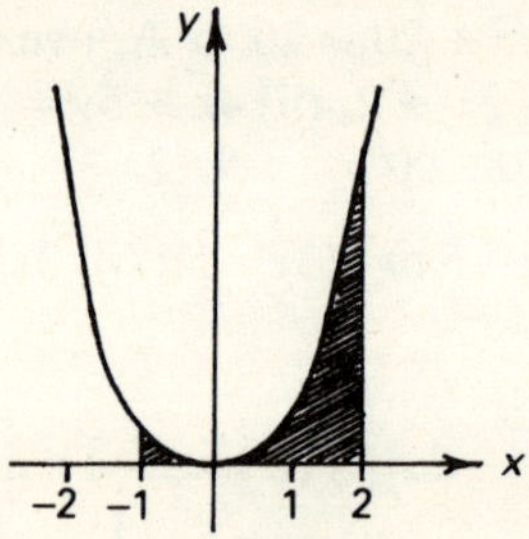

SET 44

1. $\left.\begin{array}{l} x = -4\frac{1}{2} \\ y = -4 \end{array}\right\}$ **2.** 80, 200 **3.** -3 **4.** $2^p = q$

5. $3x - y = 7$ **6.** $\frac{4}{21}$ **7.** 300

8. (i) π; (ii) π; (iii) 4π **9.** $\frac{1}{4}$ **10.** $3\sin^2 x \cos x$

SET 45

1. $a^2 + ab - ac - 2b^2 + bc$ **2.** $3 < c < 13$ **3.** 4 **4.** $\Delta = 0$

5. -1 **6.** 2 **7.** $\{0, \frac{\pi}{3}, \pi, \frac{5\pi}{3}, 2\pi\}$ **8.** $50{\cdot}5°$

9. $4\pi r + 2\pi h$ **10.** $\frac{1}{2}x^4 - \frac{1}{2}x^2 + c$

SET 46

1. $(a-c)(a-2b+c)$ **2.** 5 mm² **3.** $\frac{2}{3}, 1, -3$ **4.** 0

5. $-\frac{3}{2}$ **6.** (2, 6, 1) **7.** $\sqrt{2}$; 45

8. (i) $\frac{\sqrt{2}}{\sqrt{3}} = \frac{\sqrt{6}}{3}$; (ii) $\frac{1}{3}$ **9.** 0 **10.** $6\sin x(1 - 2\cos x)^2$

SET 47

1. $2\frac{1}{2}$
2. (i) $y = 6$; (ii) $x = 5$
3. (i)

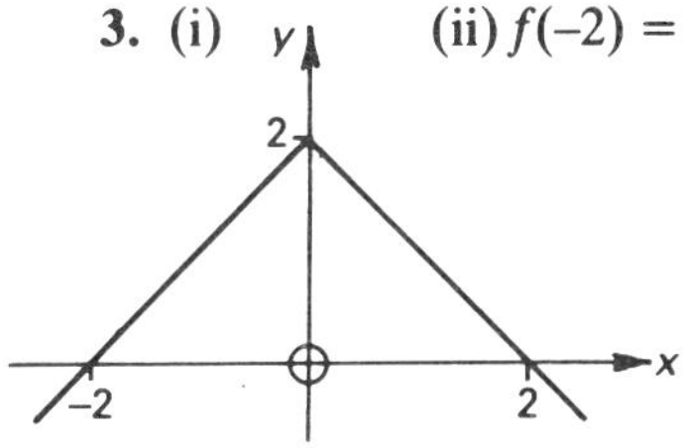

(ii) $f(-2) = 0 = f(2)$, not 1 – 1

4. $\{(0, 0), (0, 1), (1, 0)\}$
5. $y = 2x - 7$
6. $4x - 3y + 25 = 0$
7. (i) 0; (ii) 0
8. (i) 6 cm; (ii) 76·4°
9. 0 10. $-k$

SET 48

1. $\dfrac{x(x-2)}{3(3x-1)}$ 2. 1 3. –9 4. 1 5. (3, –2)
6. $a^2 - b^2$ 7. $-\sqrt{2}$; 135 8. 2 9. $x < -1$ and $x > 1$
10. $2(2x + \cos x)(x^2 + \sin x)$

SET 49

1. $h = \dfrac{3V - 2\pi r^3}{3\pi r^2}$ 2. $(3, 1\frac{1}{2})$ 3. $\cos 2x$ 4. $k \leqslant \dfrac{4}{3}$
5. $3x - 2y = 13$ 6. $x^2 + y^2 \pm 4x \pm 4y = 0$ 7. $-\sqrt{3}$
8. $\dfrac{1}{\sin x° \cos x°}\left(= \dfrac{2}{\sin 2x}\right)$ 9. 1
10. (i) $x^2 + x + c$; (ii) $\frac{1}{8}(2x + 1)^4 + c$; (iii) $\frac{1}{16}(2x + 1)^8 + c$

SET 50

1. 3 2. $\left\{\theta : \dfrac{\pi}{6} < \theta < \dfrac{5\pi}{6}\right\}$
3. $(x-1)(x+1)(x-2)(x+2)$ 4. 2·41 5. $-\dfrac{5}{7}$ 6. 90°
7. –5; 233·1
8.

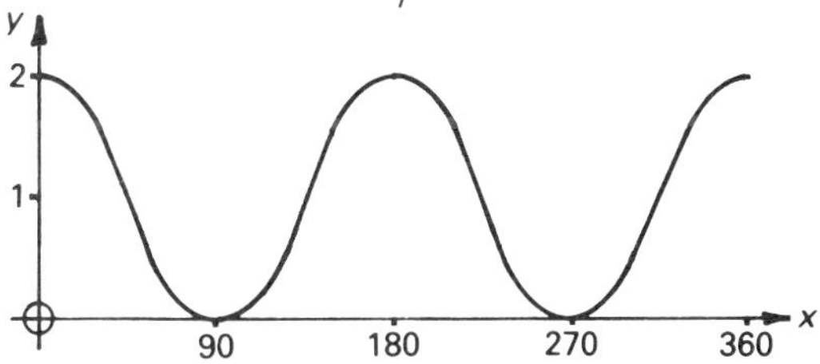

9. 3
10. $\dfrac{1-4x}{2(2x^2 - x - 6)^{3/2}}$

SET 51

1. a^{n^2-2n} 2. $\dfrac{4\pi}{3}$ cm 3. 1 4. $\triangle < 0 \Rightarrow$ unreal 5. (1, 3)
6. $4x + 3y = 25$ 7. $\dfrac{1}{7}$ 8. 73·7° 9. $\dfrac{3-2x}{2(3x - x^2)^{1/2}}$ 10. 4

SET 52

1. 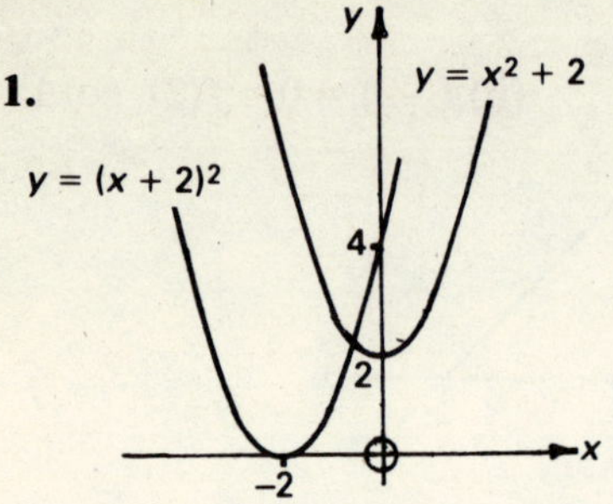

2. sin 42°; cos 48° 3. 3, 4, $2\frac{1}{2}$

4. $\log_3 6$ 5. $\frac{11}{12}$

6. $\frac{1}{4}(\underline{p} + \underline{q} - 2\underline{r})$ 7. 5; 323·1

8. 30, 150, 210, 330 9. $\frac{1}{2}$

10. $-2\cos x \sin x = -\sin 2x$

SET 53

1. $3y(3x + 2y)(2x - 5y)$ 2. 2 3. 16 4. $p^2 \geqslant 28 \Rightarrow 4p^2 - 96 \geqslant 16 > 0$

5. $(7\frac{1}{2}, -1\frac{1}{4})$ 6. 7 7. $\frac{1}{2}$ 8. 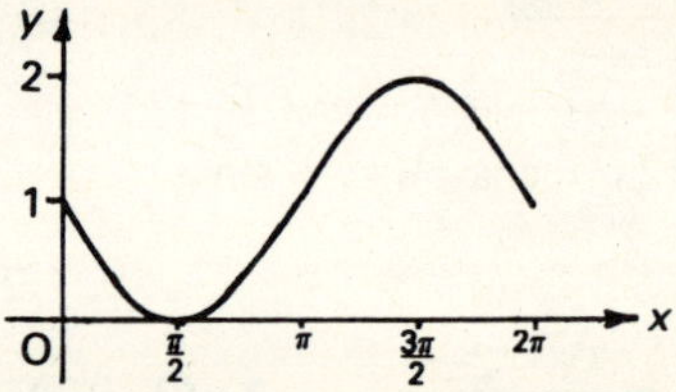

9. 2

10. $\frac{1}{2}(1 - \cos 2u)$

$\frac{1}{2}u - \frac{1}{4}\sin 2u + c$

SET 54

1. $6x^3 + 7x^2 - 23x - 30$ 2. (i) 360°; (ii) 180° 3. 1 (u_3)

4. 1·77 5. $4y = 3x$

6. 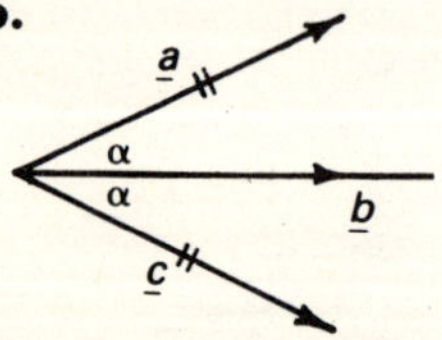

7. 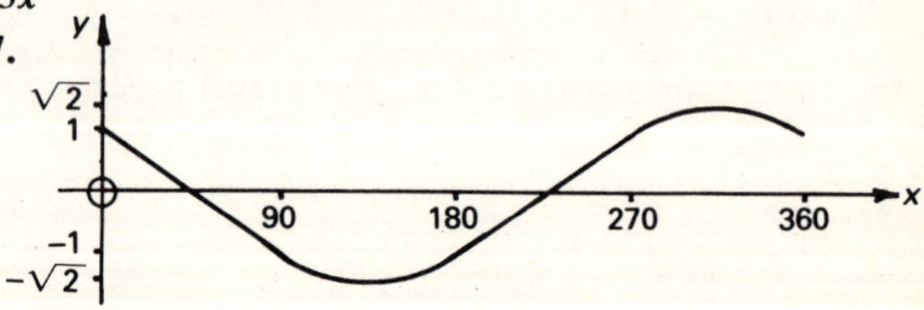

8. 84° 9. $g' \geqslant 0$ since $\cos x \geqslant -1$ 10. $2anx(ax^2 + b)^{n-1}$

SET 55

1. $6b + 7l - 42$ cm^2 2. (2, 4)

3. (i) $\{x : -3 \leqslant x \leqslant 3\}$; (ii) $\{y : 0 \leqslant y \leqslant 3\}$

4. $\{(-5, 1), (-5, 2), (-5, 3)\}$ 5. 10units^2

6. (i) $3y = 2x$; (ii) (3, 2); (iii) $x^2 + y^2 = 13$ 7. $(2\cos x - 1)(\cos x + 1)$

8. 70·5° 9. 3 10. $\int_0^a \left(g(x) - h(x)\right) dx$

SET 56

1. $-2 \leqslant k \leqslant 2$ 2. (i) 70·5, 90, 270, 289·5; (ii) 90 3. 28; 5

4. $\log \frac{x^3}{y^2}$ 5. $\sqrt{66}$ 6. $\frac{14}{15}$ 7. 12; 305

8. (i) $-\cos\theta$; (ii) $\cos\theta$; (iii) $\cos\theta$ (iv) $-\cos\theta$ 9. (0, 5) (−2, 13)

10. $-6\sin 2x \cos^2 2x$

SET 57

1. real, distinct **2** $-\frac{\sqrt{3}}{2}$ **3.** $x^2 + x$ **4.** $1 - \log_{10} 2$ **5.** -2

6. (i) 1; (ii) 0

7. $\frac{4\sqrt{3}}{7}$; $\frac{3\sqrt{3}}{14}$

8. $\frac{1}{2}$

9. $1 + \frac{1}{x^2}$

10.

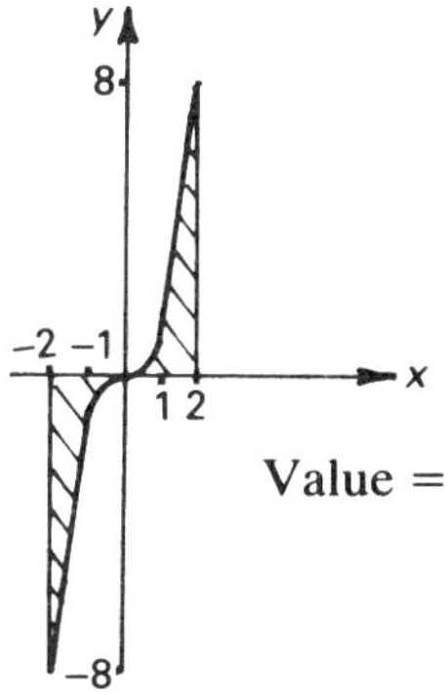

Value = 0

SET 58

1. 10 **2.** $(\cos A - 1)(\cos A - 2)$ **3.** 305

4. $a^2 + (3-k)a + (4-2k) = 0$ **5.** $(1, 2, -6)$ **6.** $a \sin\theta$

7. $\sqrt{2}$; 135 **8.** 5; -3 **9.** $-2\sin(2x+3)$

10. $\int_{-4}^{2} \left(f_2(x) - f_1(x)\right) dx$

SET 59

1. (i) $2x^2$; (ii) $x^2 + 3x$ **2.** $\{15, 165, 195, 345\}$ **3.** $9n^2$ **4.** 5

5.

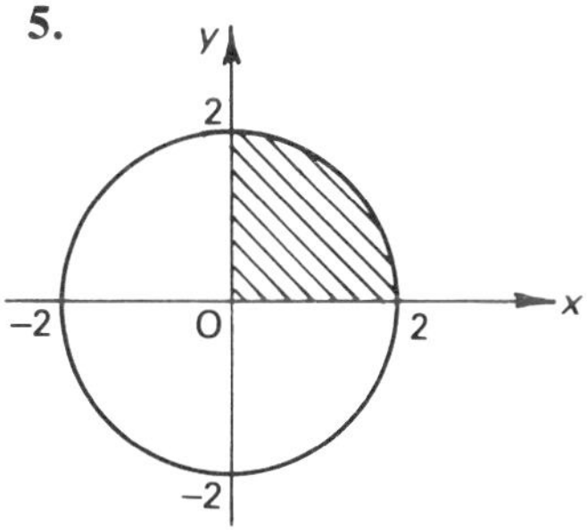

6. obtuse

7. 13 cm; 13·3°

8. $\cos 2A$

9. $-\sqrt{3}$

10. $12x(2x^2 - 1)^2$

SET 60

1. 7 **2.** (i) 25; (ii) 7 **3.** $-1, 3, 0, 16$

4. $f(0) = 1, f(1) = -4 \Rightarrow f(x) = 0$ for some x between 0 and 1

5. $(x-2)^2 + (y-1)^2 + (z-3)^2 = 25$ **6.** $\overrightarrow{XY} = 3\overrightarrow{YZ}$ **7.** 2π

8. $x \sin x$ **9.** $4x^2 + 6x + 2$ **10.** $\cos x \cos(1 + \sin x)$

SET 61

1. 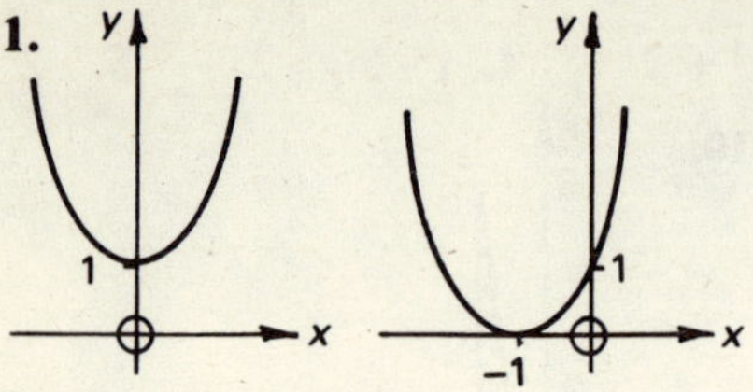

2. $x = h$ 3. $x \leqslant 6$

4. $f(p) = 0$

5. $h^2 + k^2 < r^2$

6. $a_1b_1 + a_2b_2 + a_3b_3$

7. (i) $\cos A$; (ii) $-\cos A$; (iii) $\sin A$; (iv) $\sin A$

8. $2 \cos x$

9. e.g. 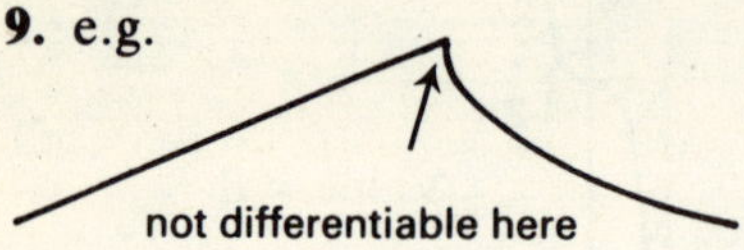

10. 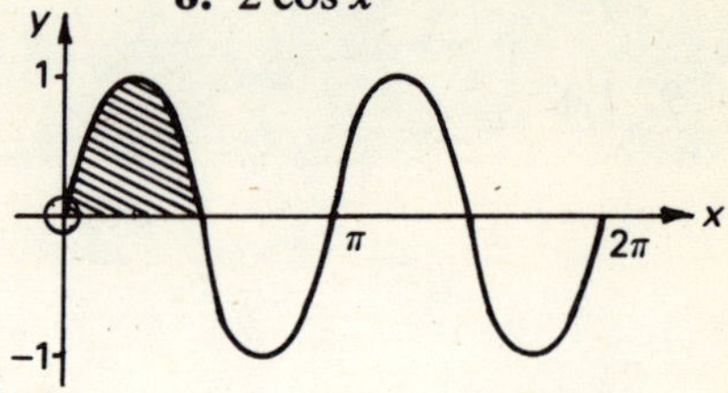

SET 62

1. (i) $\dfrac{1}{2x}$ (ii) $\dfrac{2}{x}$ 2. (i) $\sin^2 A$ (ii) $\cos^2 A$

3. $b = a^x$ 4. $x \neq k, x \neq 1$ 5. Parallelogram

6. $\sqrt{(x_2 - x_1)^2 + (y_2 - y_1)^2 + (z_2 - z_1)^2}$

7. (i) $\frac{1}{2}(1 + \cos 2A)$ (ii) $\frac{1}{2}(1 - \cos 2A)$

8. $\sqrt{2} \cos\left(x - \dfrac{\pi}{4}\right)$ 9. 1 10. $\mathrm{F}(b) - \mathrm{F}(a)$

SET 63

1. $k = \dfrac{x\sqrt{z}}{y^3}$ 2. (b, a)

3. If a splog is crunky then it is dargy, or All crunky splogs are dargy.

4. $q^2 - 4pr < 0$ 5. $\sqrt{a^2 + b^2 + c^2}$

6. (i) $\left(\dfrac{r_1 + s_1}{2}, \dfrac{r_2 + s_2}{2}, \dfrac{r_3 + s_3}{2}\right)$ (ii) $r_1s_1 + r_2s_2 + r_3s_3$

7. $\dfrac{5}{4}$ 8. (i) $\sin 2x$; (ii) π 9. $10 \sin 5x \cos 5x$ $(= 5 \sin 10x)$

10. $\dfrac{x}{2} - \dfrac{1}{20} \sin 10x + c$

SET 64

1. $y = \dfrac{kx^2}{z}$ 2. (i) 1; (ii) 1; (iii) 1

3. 2 4. $\log_a b = x$

5. $\left(\dfrac{a}{2}, \dfrac{b}{2}\right)$ 6. collinear

7. (i) 1; (ii) $-\cos 2A$; (iii) $\sin 2\theta$

8. 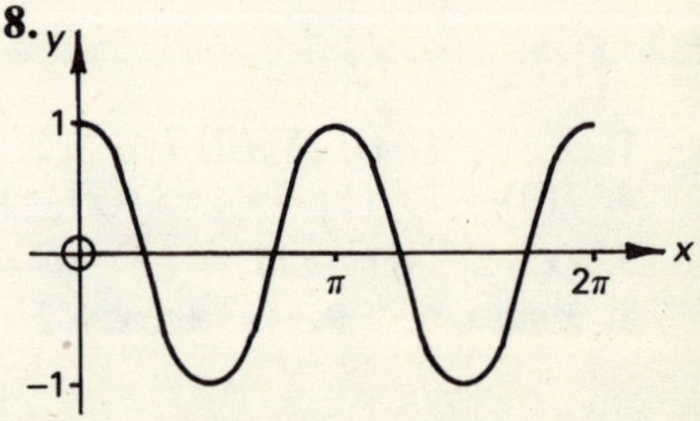

9. $-\sin x + \cos x$ 10. $k + l$

SET 65

1. 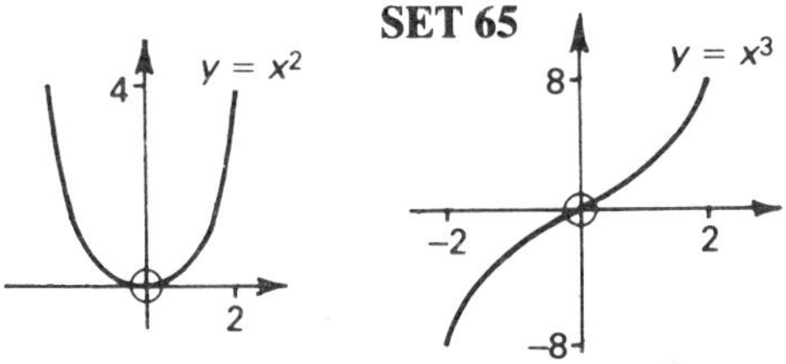

2. $x = k$ 3. (i) 1; (ii) 1; (iii) 0 4. $y = x^2 + x$ 5. 1, 0

6. $\dfrac{y_2 - y_1}{x_2 - x_1}$ 7. (i) $\cos(A + B)$; (ii) $\sin(Y - X)$

8. (i) 2; (ii) 3; (iii) 1 9. $f'(a)$ 10. 0

SET 66

1. $x = \dfrac{-k \pm \sqrt{k^2 - 4hl}}{2h}$ 2. $\left(\dfrac{x_a + x_b}{2}, \dfrac{y_a + y_b}{2}\right)$

3. (i) $a^{-n} = \dfrac{1}{a^n}$; (ii) $a^{1/n} = \sqrt[n]{a}$; (iii) $\log_a X - \log_a Y = \log_a \dfrac{X}{Y}$;
(iv) $\log_b XY = \log_b X + \log_b Y$

4. 3^{n-1} 5. $y - k = n(x - h)$ 6. $\underline{p} = \frac{1}{3}(\underline{m} + 2\underline{n})$

7. (i) $-2 \sin A \sin B$; (ii) $-2 \cos A \sin B$

8. (i) π; (ii) $\dfrac{\pi}{2}$; (iii) $\dfrac{2\pi}{3}$; (iv) 4π; (v) 8π

9. (i) $\cos x$; (ii) $-\cos x + c$ 10. $\displaystyle\int_a^b g(x)\, dx$

SET 67

1. $x = 1$ 2. (i) $-\cos A$; (ii) $\cos A$ 3. $\log y = n \log x + \log a$

4. (i) $\dfrac{\sqrt{a}}{a}$; (ii) $\dfrac{\sqrt{a} + b}{a - b^2}$ 5. $-\dfrac{1}{m}$

6. $\dfrac{a}{d} = \dfrac{b}{e} = \dfrac{c}{f}$ or $a = kd, b = ke, c = kf$

7. 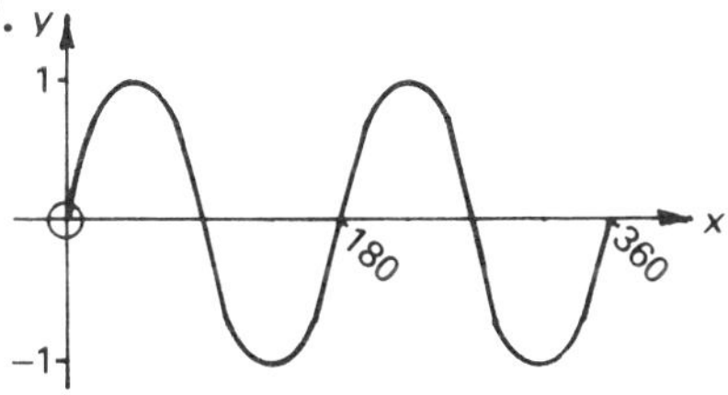

8. 5

9. $f'(x) > 0$

10. $-\dfrac{1}{2}\cos 2x + c$

SET 68

1. 3 2. Triangular pyramid with all edges equal

3. $\dfrac{5}{3}$ 4. $ah^2 + bh + c$ 5. $c = 0$ 6. 0

7. (i) $2\cos^2 A - 1$; (ii) $2\cos^2 2A - 1$; (iii) $1 - 2\sin^2 \dfrac{A}{2}$

8. (i) 1; (ii) 2; (iii) 2 9. $f'(b)$ 10. $2k$

SET 69

1. (0, 1) **2.** Angle subtended at centre of circle by arc equal to radius
3. $\log_a XYZ$ **4.** ar^n **5.** $(x-u)^2 + (y-v)^2 = r^2$
6. (i) scalar; (ii) vector; (iii) vector **7.** (i) 1; (ii) $\cos x$; (iii) $\sin x$
8. (i) 2π; (ii) π; (iii) $\frac{2\pi}{3}$ **9.** $na(ax+b)^{n-1}$ **10.** $\int_a^c f(x)\,dx$

SET 70

1. $x^3 - 6x^2 + 8x + 4$ **2.** $\frac{1}{2}qr \sin P$ **3.** No (not 1 – 1)
4. $f(x) = kx^2(x-4)$; $k = -1$; $\Rightarrow f(x) = 4x^2 - x^3$
5. (i) a line from A to the mid-point of BC; (ii) a line from B, $\perp^r$ to AC
6. (i) 2; (ii) 6; **7.** 1 **8.** (i) $\frac{1}{2}$; (ii) -1 **9.** $f'(a) = 0$
10. (i) $-\frac{1}{2}\cos 2x + c$; (ii) $\frac{1}{2}\sin 2x + c$

SET 71

1. (i) 2; (ii) 1 **2.** $y = mx + 4$ **3.** $\Delta \geqslant 0$
4. (i) $\log_a \frac{1}{n} = -\log_a n$; (ii) $\log_a X^n = n \log_a X$
5. $(-h, -k)$, $\sqrt{h^2 + k^2 - l}$ **6.** $|\underline{u}| = 0$, $|\underline{v}| = 0$, $\underline{u} \perp \underline{v}$
7. yes, it has length 1 **8.** $b > a$
9. rate of change of f with respect to x or gradient of tangent to graph of f at $x = 1$
10. (i) $\frac{(ax+b)^{n+1}}{a(n+1)} + c$; (ii) $-\frac{1}{a}\cos(ax+b) + c$

SET 72

1. $0 \leqslant 3x \leqslant 6x$ **2.** $\frac{b^2 + c^2 - a^2}{2bc}$ **3.** All men are bus drivers
4. $P = kQ^a$ **5.** They are perpendicular **6.** $\frac{\underline{a} \cdot \underline{b}}{|\underline{a}||\underline{b}|}$
7. (i) 2π; (ii) 2π; (iii) 2π **8.** 60, 120, 240, 300
9. $f'(a) = 0$ **10.** (i) $-x + c$; (ii) $-\sin x + c$

SET 73

1. (*f*) **2.** (*b*) **3.** (*j*) **4.** (*h*) **5.** (*l*)
6. (*g*) **7.** (*c*) **8.** (*a*) **9.** (*k*) **10.** (*e*)

SET 74

1. F **2.** F **3.** F **4.** F **5.** F
6. F **7.** F **8.** T **9.** T **10.** T

SET 75

1. 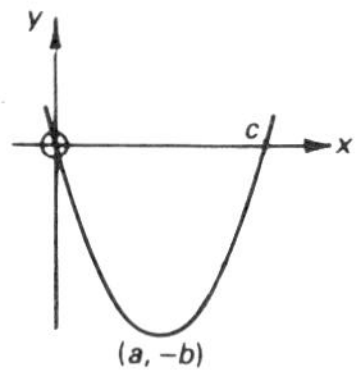

2. 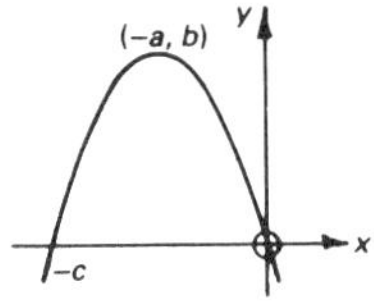

3. 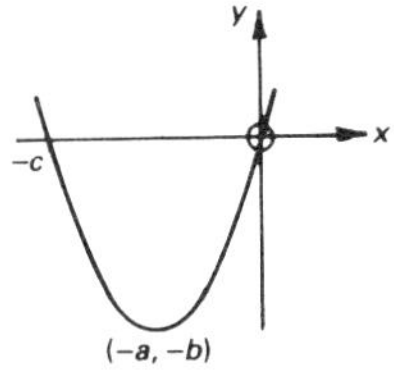

4. 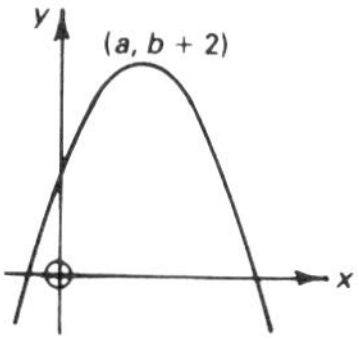

5. 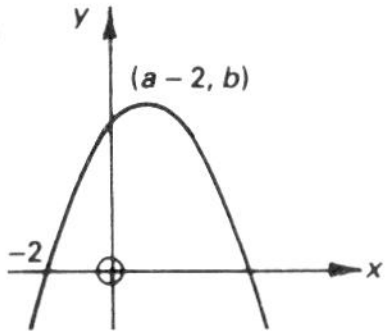

6. 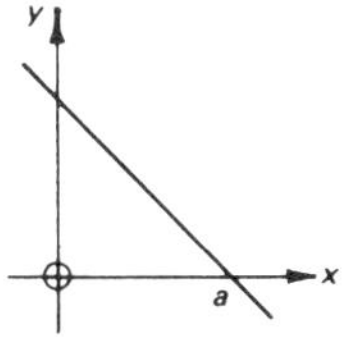

7. 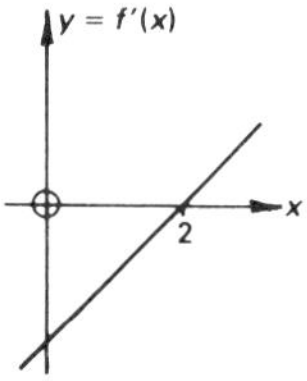

8. 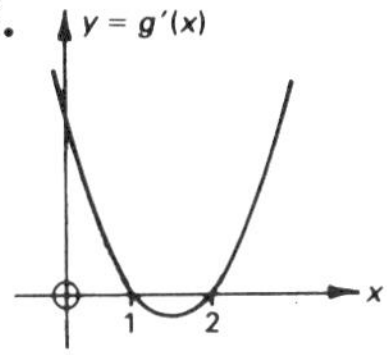

9. 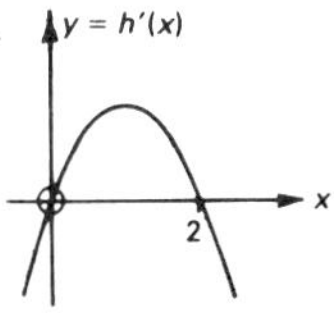

10. 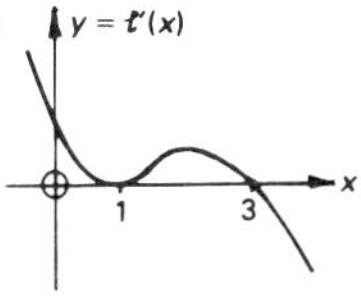

SET 76

1. 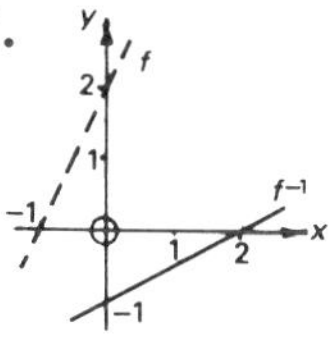

2. 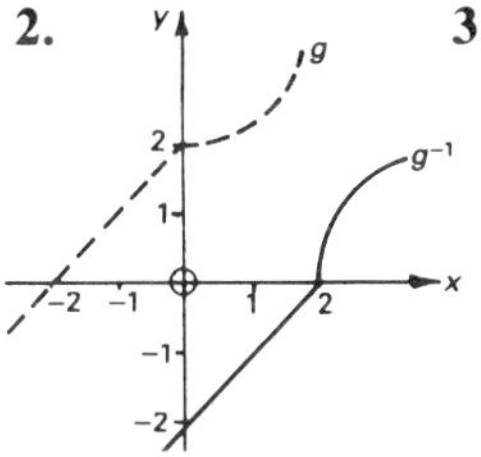

3. 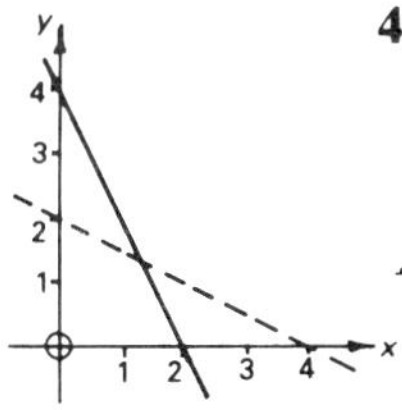

4. 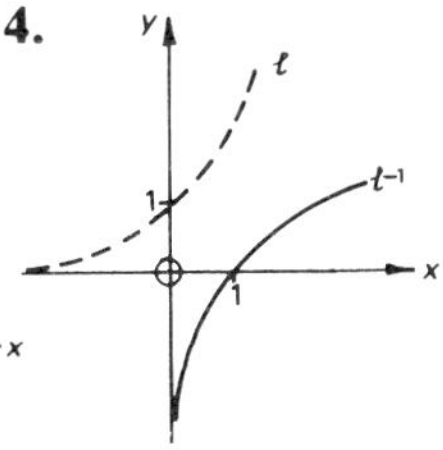

5. $f(x) + 2$ **6.** $-f(x)$ **7.** $f'(x)$ **8.** $f(-x)$ **9.** $f(x + 2)$ **10.** $-f(-x)$

SET 77

1. 99 **2.** 999 999 **3.** *(a)* 6; *(b)* 5 **4.** 3 **5.** 80

6. 11 **7.** 36 cm^2 **8.** $\dfrac{5}{16}$

9. $\left(\text{Hint } a^2 - 2.\pi r^2 \text{ with } r = \dfrac{\sqrt{2}\,a}{4}\right)$ gives $a^2 - 2\pi\left(\dfrac{\sqrt{2}\,a}{4}\right)^2$

10. $a \,.\, 1000 + b \,.\, 100 + b \,.\, 10 + a = (1001)a + (110)b = 11(91a + 10b)$

RECORD YOUR PROGRESS ON THIS CHART

Date	Set	Done	Score	Date	Set	Done	Score	Date	Set	Done	Score
	1				27				53		
	2				28				54		
	3				29				55		
	4				30				56		
	5				31				57		
	6				32				58		
	7				33				59		
	8				34				60		
	9				35				61		
	10				36				62		
	11				37				63		
	12				38				64		
	13				39				65		
	14				40				66		
	15				41				67		
	16				42				68		
	17				43				69		
	18				44				70		
	19				45				71		
	20				46				72		
	21				47				73		
	22				48				74		
	23				49				75		
	24				50				76		
	25				51				77		
	26				52						

Printed in Great Britain by
Martin's of Berwick Ltd.

THREE SUSSEX POETS

Russ Bravo
Steve Carroll
Martin Collins

WHITE DOOR PRESS

First published in the United Kingdom in 2009
by White Door Press, 27a Goring Road, Worthing,
West Sussex BN12 4AR.

Design: www.inspiration-by-design.co.uk

Printed in England by Colorwise, Worthing.

RUSS BRAVO

Russ is a journalist and writer whose love of words emerges in poetry, songs and comedy on stage, radio and TV. He is Editor of *Inspire* magazine and runs Matt's Comedy Club in Worthing.

HANDS UP

HANDS UP IF YOU FEEL OLD
if your golf shoes are full of mould
if your sell-by date has just been sold - hands up

hands up if you feel tired
if you're not a taste that's easily acquired
if there's not one oil painting you've inspired - hands up

hands up if you'd quite like a sleep
if your life's treading water and the water's too deep
if you've ever thought it must be quite nice to be a sheep - hands up

hands up if your knees often creak
if you'd like to give your MP's nose a tweak
if you've ever wondered if the Eskimo ever does a streak - hands up

hands up if you've joined the crew
hands up if you've tamed the shrew
hands up if you don't care you're not in Who's Who
hands up

COOKS

Too MANY COOKS SPOIL THE BROTH
it is said,
In which case
perhaps some cooks should consider
doing something else
And leave the soup
To those who are s(o)uper
At it.

STRING

How long is a piece of string?
It depends on the string,
but most pieces are as long
as one end to another.
So next time you ask
someone a question, and
they answer: "How long is a
piece of string?" (meaning they don't know)
Just get out your prepared piece
of string, and reply: "21 cm*".
(If it is)

* *Adjust according to length*

WALK TALL

THERE WAS AN OLD SONG
that went: "Walk tall,
and look the world right
in the eye".
But you can only really walk
as tall as you are, so looking
the world in the eye really depends
on whether the world is at (your) eye-level.
You're as tall as you are, generally.
But actually, you're either growing, in which case
you're not as tall as you will be.
Or you've stopped growing, and have probably
started shrinking, in which case you're not as
short as you're going to end up.
Which is a worry.
So I'd adjust the song to read:
"Walk upright, and look at the world in whatever
way suits your height best"
And you may want to invest in some stilts.

HOME BAGGAGE

A SNAIL
Carries his home upon his back
A hiker
Carries his rucksack upon his back
But is less slimy.
Usually.

SKY PAINTER

The sky is a canvas
On which God paints
In cloud.
Quite often there is blue
Depending on where you live.
And there can be black, grey and white.
And with a glorious flourish
You will often find gold and pink.
None are kept - for God is creative
Every day.
Unlike some artists, who need inspiration.

STRANGE

I FIND IT STRANGE
That when grass grows
It grows up straight
Not along, or in curls.
Imagine if our hair did that.
Rather than hanging down
Or getting thicker
It just got taller
So that eventually
whoever cut your hair
Would need a stepladder
And very tall ceilings.
And highlights
Would mean something else
Altogether.

STRENGTH IN NUMBERS

There is strength in numbers
They say.
Not quite understanding what they meant
I considered which numbers might be strongest.
Three I'd always thought quite tough
But six seemed feeble, somehow.
Ten had something going for it
But 13 was always considered unlucky
And therefore unreliable.
Seven was good and solid
But four could cause problems.
Nine was always wanting to be ten
I suspected.
A million sounded tough as iron
- but what if it was all made up of fours
and sixes?
Would lots of weebly numbers put together be strong?
Or just extra weebly?
Then I discovered they weren't talking about
numbers at all, but people.
Which was a bit annoying, really.
If they'd said: "There is strength in people being together"
It would have been much simpler.
Unless there were only six of them, maybe.

THE DAY BAD LUCK RAN OUT

I walked on the cracks in the pavement
But the lions didn't get me
I walked under a ladder
But the paint didn't wet me
I dropped a mirror and broke it
But bad luck seemed to forget me
A black cat crossed my path
But that just didn't fret me
On Friday the 13th
Even school didn't deject me

Maybe superstition just gave up
And decided to respect me.

BODY BITS

I'M GLAD THERE'S A BUTTON ON MY TUMMY
And lids on both my eyes
I'm happy there are caps upon each knee
And they're both about the same size

I'm delighted with my nostrils
There's one each side of my nose
And I'm chuffed both my arms have elbows
(For my funny bones, I suppose)

I'm tickled that my eyes have lashes
And each of my teeth has a gum
I'm well pleased both feet have an ankle
And my hips (at the back) have a bum
Both of my ears boast a lobe, you know
And my ribs are kept in a cage
My knuckles help keep my fingers on
It's probably down to my age

WHAT SHALL WE CALL THIS POEM?

What shall we call this poem?
What is it really about?
Is it short, is it fat, is it long, is it thin?
Its subject is really in doubt.

We could just write about nothing.
But that would be quite boring.
If we wrote about nothing,
Then you'd soon be snoring.

Maybe we'll write about chickens.
Maybe we'll write about toast.
Perhaps we'll scribble about monkeys
That picnic by the coast.
Is there a pig in your toilet?
Have your aunties got fleas?
Does your hamster always do handstands?
Then sleep for hours dreaming about cheese?

There may be no end to this poem
It just rambles endlessly on
So we'll just put it out of its misery
And stop.

(Now it's gone)

PALE

THE WIND HAS PAINTED MY TRACTOR
Several shades of tree
The walrus in my garage
Has just turned 43
A walnut phoned me up to say
The crisps are tasting stale
And still my auntie says to me:
"My, you're looking pale ..."

The badgers living on our roof
Have stolen all the scones
A bison with a dodgy leg
Just scored for MK Dons
The blancmange in our cupboard
Is now shaped just like a whale
No wonder my auntie says to me:
"My, you're looking pale ..."

I thought that everyone would find
A volcano in their fridge
And be woken by the vultures
Every night, as they play bridge
Me, I blame the Vimto-covered
Hippo, built to scale
It's his fault my auntie says to me:
"My, you're looking pale ..."

VALENTINE DILEMMA

I did not want to
Buy you something
That put you in a pigeon hole
Or stereotyped you.
Or made a joke at your expense.
That you might not understand
Or see the spirit in which the joke was made.

So that ruled out:

Roses
Anything pink
Anything with teddy bears on it
Or indeed, teddy bears themselves.

Cushions in the shape of hearts
Or lips
Anything red, maybe
Or satin
Or lingerie
Or chocolates. (because that's so obvious)
Maybe.

Cupid would be stupid
And who was St Valentine anyway?

So I bought you
A set of spanners …
Because you're practical and they're useful
And you're better at repairs and plumbing
Than me.

And I scratched some kisses on one of them
So whenever you're tightening a nut
You'll think of me.

That's a bit romantic, isn't it?

STRANGE FISH

You can't see
You can't see the beach for the wood
For the wood has become forest
Again
Planks stretched aloft
Like the yearning arms of the crowd at the 100 Club
In '76, wanting to catch the salt drops of rebellion
Before they become the dried sweat of the ordinary
Pale pine fingers crossed
In hope of rescue
'MAN THE LIFEBOATS
WOOD OVERBOARD'
Cargo gone
Benches pressed
Decking bobbing
Shelving diving
And suddenly arriving
An angular invasion force
A splinter group that doesn't mind being split
Or savaged with steel teeth. Or screwed. Or nailed.
Elvis would have liked our beach.
It once had a wooden heart.

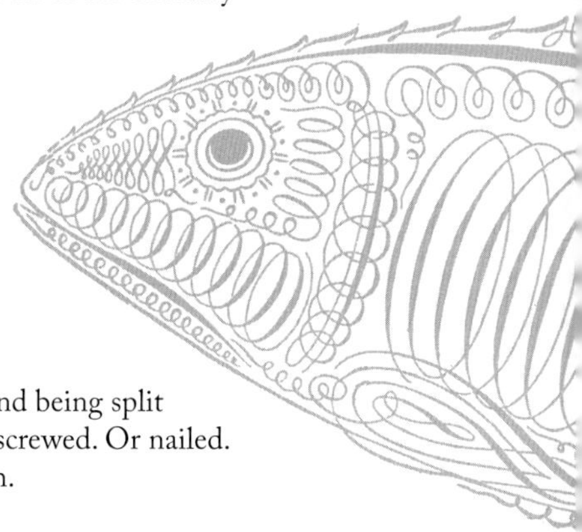

This was written for the 2009 Ice Prince arts festival - marking one year since Worthing's wood invasion, due to the discharged cargo out at sea.

THE SHELF STACKERS

I MUST GO DOWN TO THE SEA AGAIN
To the roaring sea and the tide
I've got 2000 tons of wood down there
I wonder if it's dried?
It's got to be someone's decking
It's bound to be someone's shelf
Have you seen the twilight trekking
Loading roofracks by stealth?
My dream conserv-a-tory
Is stacked up on the shingle
The DIY man's glory
The thought just makes him tingle
With some sawing and some planing
With his drill and Black & Decker
Some awesome light construction
Hardly make him a villainous wrecker
While his joints won't be mortice and tenon
More Superglue and bodge
And his angles may look a bit wonky
More tumbledown shed than park lodge
It'll be fashioned from flotsam and jetsam
The stuff that the tide just brought in
Like a boon from above, a woodworker's offering of love
Strewn from the god of Homebase's bin

You should have seen their cars go
Weighed down with the loot
Half a dozen splintered planks
Protruding from the boot
It may not be strictly legal
But many a blind eye was turned
And many's a glass has been raised to the Prince
As some flaming good firewood's been burned.
Nice one, Icey.

Another poem written for the 2009 Ice Prince arts festival.

STEVE CARROLL

Steve is an artist and graphic novelist. He is the author of the *Riddler's Fayre* series.

I WANT TO BE YOUR ENEMY

I WANT TO BE YOUR ENEMY,
Don't care for what you say,
I think I'll make my own mind up,
it's me who has to pay,
the price for all my errors,
the cost for all my thoughts
And how I could be making things,
while adding up your noughts.

I want to be your enemy,
I want you to despair
at how I'm walking backwards
with my hands up in the air,
at how I still don't get it
and want you to explain
when the chances don't make dances
again and again and again.

I want to be your enemy,
I don't want to be your friend,
I want to get inside your head
and drive you round the bend.
I want you to remember
I don't fly beneath your wing,
That I have my own rebel song,
so listen to me sing:

All things bright and beautiful,
All creatures great and small,
All things wise and wonderful,
The Lord God made them all.

104 MILLION DOLLARS

I HAD 104 MILLION DOLLARS,
So I asked what I could buy:
A multi storey building,
Not quite scraping the sky.
I could be a saint
And build 500 brand new churches,
Or cover Hampstead Heath
With 2 million silver birches.
Or 20 million rounds of ammo
For as many holes in someone's head.
Or a painting …
by Picasso … instead.

Twelve hundred BMWs
With satellite navigation,
The total annual budget
Of a developing Third World Nation;
The reward for 6,000 apostates
Hanging by their necks;
Six hundred thousand Asian kids
Sold to me for sex.
I could bribe 48 thousand judges
To say I wasn't there at all;
Or buy a painting by Picasso
To hang … on me wall.

Three hundred million plastic dinosaurs
that go SQUEEK! SQUEEK! SQUEEK!
Or I could feed the whole of
the Sudan ... for a week.
Cover all of Coventry
In a fabric that never fades;
Or therapy for 2 million
Africans with AIDs;
Re-house 30 million Iraqis
Because it never was their fault;
Nah! I'll buy a painting by Picasso
And I'll keep it ... in a vault.

A SOMBRE HYMN SUNG TO NO-ONE

On a dark cold night in flight
A cathedral town by candle-light
The stained glass replaced with clear rational
A secular choir in the cathedral mall.
Promising peace their praises run
Like a somber hymn sung to no-one,
And the songs rise to bilious night
Like an anxious dream at twilight.

On the forest crest a blood-red dawn
With no future but dying and porn
And in the half light two men fight
And both claim to walk by sight
But one man says we rule the heart
The other claims we fall apart
As the landfill earth reflects the sun
In a somber hymn sung to no-one.

Gone is the goose with the golden rule
No longer sung the melody of fools
In the red brick building all history has become
Just a somber hymn sung to no-one
And the two men die in each other's grasp
When a small child does timidly ask:
"Why do we bother to curse the night
when nothing is wrong and no-one is right?"

CREED AT 44

I FIND IT HARD TO BELIEVE IN GOD,
I find it hard to lose weight,
I find it hard to say no
to bargain bookshops
and cheap CDs
I find it hard to stay awake,
during Big Brother and "100 Best…"
and not to resist taking the coastal road
on the way home.
I find it hard to discipline children
who say rude jokes, that make me laugh
I find it hard to grow old.

But I don't find it hard
to grow grass on the ground
where children challenge me to chase them
or follow the flow of a historians hype.
I don't find it hard to love language
or be moved by a miniature in the corner of a museum'
or look upon life as learning.
I don't find it hard
to eat when I'm hungry
or breath when I'm breathless
or cry when I'm happy
So I don't find it hard
to believe in God.

POPULATION EXPLOSION

ANG! BANG! BANG! BANG! BANG! BAN
ANG! BANG! BANG! BANG! BANG! BAN
ANG! BANG! BANG! BANG! BANG! BAN
ANG! BANG! BANG! BANG! BANG! BAN
ANG! BANG! BANG! BANG! BANG! BAN
ANG! BANG! BANG! BANG! BANG! BAN
ANG! BANG! BANG! BANG! BANG! BAN
ANG! BANG! BANG! BANG! BANG! BAN
ANG! BANG! BANG! BANG! BANG! BAN
ANG! BANG! BANG! BANG! BANG! BAN
ANG! BANG! BANG! BANG! BANG! BAN
ANG! BANG! BANG! BANG! BANG! BAN
ANG! BANG! BANG! BANG! BANG! BAN
ANG! BANG! BANG! BANG! BANG! BAN
ANG! BANG! BANG! BANG! BANG! BAN
ANG! BANG! BANG! BANG! BANG! BAN
ANG! BANG! BANG! BANG! BANG! BAN
ANG! BANG! BANG! BANG! BANG! BAN
ANG! BANG! BANG! BANG! BANG! BAN

ANG! BANG! BANG! BANG! BANG! BAN
ANG! BANG! BANG! BANG! BANG! BAN
ANG! BANG! BANG! BANG! BANG! BAN
ANG! BANG! BANG! BANG! BANG! BAN
ANG! BANG! BANG! BANG! BANG! BAN
ANG! BANG! BANG! BANG! BANG! BAN
ANG! BANG! BANG! BANG! BANG! BAN
ANG! BANG! BANG! BANG! BANG! BAN
ANG! BANG! BANG! BANG! BANG! BAN
ANG! BANG! BANG! BANG! BANG! BAN
ANG! BANG! BANG! BANG! BANG! BAN
ANG! BANG! BANG! BANG! BANG! BAN
ANG! BANG! BANG! BANG! BANG! BAN
ANG! BANG! BANG! BANG! BANG! BAN
ANG! BANG! BANG! BANG! BANG! BAN
ANG! BANG! BANG! BANG! BANG! BAN
ANG! BANG! BANG! BANG! BANG! BAN
ANG! BANG! BANG! BANG! BANG! BAN
ANG! BANG! BANG! BANG! BANG! BAN
ANG! BANG! BANG! BANG! BANG! BAN

ANG! BANG! BANG! BANG! BANG! BAN
ANG! BANG! BANG! BANG! BANG! BAN
ANG! BANG! BANG! BANG! BANG! BAN
ANG! BANG! BANG! BANG! BANG! BAN
ANG! BANG! BANG! BANG! BANG! BAN
ANG! BANG! BANG! BANG! BANG! BAN
ANG! BANG! BANG! BANG! BANG! BAN
ANG! BANG! BANG! BANG! BANG! BAN
ANG! BANG! BANG! BANG! BANG! BAN
ANG! BANG! BANG! BANG! BANG! BAN
ANG! BANG! BANG! BANG! BANG! BAN
ANG! BANG! BANG! BANG! BANG! BAN
ANG! BANG! BANG! BANG! BANG! BAN
ANG! BANG! BANG! BANG! BANG! BAN
ANG! BANG! BANG! BANG! BANG! BAN
ANG! BANG! BANG! BANG! BANG! BAN
ANG! BANG! BANG! BANG! BANG! BAN
ANG! BANG! BANG! BANG! BANG! BAN
ANG! BANG! BANG! BANG! BANG! BAN
ANG! BANG! BANG! BANG! BANG! BAN

SEASONAL BELIEVER

No more suncream factor 5
No more outdoor music live
No more grass stains on your jeans
No more recreation machines
No more sleep beneath thin sheets
No more parasols in the streets
No more barbecue burgers frying
It is Autumn and your god is dying.

Hello to hot milk at dawn
And dried blood leaves across your lawn
Hello to repeats on TV
(Hello to phototherapy)
Hello to vows of staying sober
With Christmas starting in October
Hello to the depressed poet sighing
It is Autumn and your god is dying.

Remember May's translucent sky
With the yearning of the year's eye?
Remember you crucified your hate
And walked with me through Easter's gate?
Remember I asked you what was wrong
In the twilight of the blackbird's song?
And I remember how you shook your head
Well, it's Winter now and your god is dead.

LOVE ME, I AM THE VOID

Love me, I am the void;
A philosophical abstraction made awesome,
A scientific implausibility,
A poetic construction, of course,
for should I speak I would sound like this ....
....
....

So, love me, I am the void,
I am the hole in your Polo mint,
The ohm in your mantra,
The aperture in your camera,
(Through which you observe)
The rift in your family.
The shaft for your axel,
The zero to activate your alphabet,
The black hole at the centre of your star system,
The very place you cannot exist.

So, love me, I am the void,
The cut-price, bargain basement
alternative to God,
with no cross to bear,
and no commandments
(apart from physics),
"Come take my vacuum,
for it is light!"

So, love me, I am the void,
You have nothing to lose when you argue
For I am the nothing you declare.
I am your best answer
And your worst nightmare.
So, come, embrace me,
I am your first cause
and your final destination.

For Richard Dawkins

EAT THIS!

It may sound odd, but I think of life;
As a delicious, crunchy thing.

You bite through the crust of small-talk and slander;
And savour the smell, as a child chuckles,
The first aroma of an undiscovered city, or
The fondness of a new found friend.

As you drink the dark wine of the blackbird's song,
Or a poem that perhaps, in a moment, pierces your perspective;
When you sink your teeth into a symphony by Schubert,
And taste the tangled forest of a troubled temperament.

It may sound odd, but I think of life;
As a delicious, crunchy thing.

And there is the anticipation of oil on canvass,
A bath full of lavender, a garden full of apples,
An eye full of seeing, but never enough!
(And never forget the sauce of sex).

I crave the ever-replenishing sea, and
The beach that sermonizes the scraps of
Many millennia of momentary meals,
The crumbs of a body, broken for you.

It may sound odd, but I think of life;
As a delicious, crunchy thing.

Some settle for the sludge of MacDoubt;
MacOught, MacCan't, MacSame-as-yesterday;
But bring me the broth of boiled history,
And I'll scoop up those croutons with a cage.

For time is the menu of opportunity,
Just ordering 'a life' is all we ought to do;
Since silence is not saintliness,
And death … is a dessert … that deliberates … daily.

It may sound odd, but I think of life;
As a delicious, crunchy thing.

HEROES CRY

The semtex travellers cheques can't compete,
With the bigoted neighbour down your street.
And the night-sight armalite can't hold a light
to the argument with the ex last night.

'Cos heroes weep and heroes cry,
not trained to ask the questions why,
And heroes bleed and heroes pray,
Beg Christ to wash their sins away.

The sandwich in the Saracen at dawn,
Was as great gift as your first-born.
The knowing that in London, Whitehall's powers,
Would let you live another 24-hours.

'Cos heroes weep and heroes cry,
not trained to ask the questions why,
And heroes bleed and heroes pray,
Beg Christ to wash their sins away.

Do you remember the look of surprise,
In the face of the Arab with the baby eyes?
As he reached for his gun and his mind grew dim,
Look, John, it was either you or him.

'Cos heroes weep and heroes cry,
not trained to ask the questions why,
And heroes bleed and heroes pray,
Beg Christ to wash their sins away.

Now you're pensioned off and your story's done,
Memories fade like the front page of The Sun.
See Action Man in the hands of a boy,
Last year's hero is this year's toy.

'Cos heroes weep and heroes cry,
not trained to ask the questions why,
And heroes bleed and heroes pray,
Beg Christ to wash their sins away.

For John

TO AN IRAQI CHILD

She was born
In the wrong place,
With the wrong god,
With the wrong face.
At the wrong time,
With the wrong tongue,
Under the wrong leader,
And the wrong sun.
With the wrong oil.
And the wrong sand,
With the wrong poets,
In the wrong land.
To the wrong Mum,
In the wrong bed,
To the wrong Dad,
And what he said.
With the wrong flag,
And the wrong friends,
With the wrong toys,
And the wrong ends.
With the wrong laugh,
And the wrong smile,
At the wrong end,
Of a cruise missile.

She must be punished for making
so many mistakes.

THE DREAMBOAT (A LULLABY)

Climb aboard, my darling boys
and let daddy wrap you warm
in a sail made from comfort and adventure,
as I push you out into the night's cold sea
I say a prayer.

I say a prayer because
you carry a history, not yet ready
to raise its head.

I say a prayer because
out at sea
are men who would exploit you
as they were exploited
and women who will hurt you
as they were hurt.

I say a prayer because
you have pearls to be revered
and joy to inhabit
and light to let loose
and belief that should not be disbelieved.

And I say a prayer because
we all need more
than just a boat filled with dreams.

For Aidan and Declan

SHUT UP!

How did Germany fall under Nazi domination?
They seemed to me such a sophisticated nation;
And how come Stalin turned the October Revolution
Into his own version of the Final Solution?
Fortunately now the only Iron Curtain
Is between the apathetic and the all-certain.
But these minor squabbles are about religion;
Serves them right for taking that position.

So crucify the Christians and bring back the lions;
Round up the Jews and send them off to Zion;
Pour petrol on the Buddhists and set them all alight;
But don't go near Muslims, they may put up a fight.

Belief is about thought not races;
Even the Simpsons have yellow faces.
And if logic lacks love, it also lacks choice;
Or the courage to hear another man's voice.
So come gather round and tell us your stories;
THIS MEETING IS OPEN TO SIKHS, HINDUS
AND TORIES!
As Clint Eastwood said as he split open rocks:
"Faith ain't worth spit if you can't take the knocks."

So laugh at the Christian as she fights off the lion;
Round up the Jews and send them off to Zion;
Take those baldy Buddhists and give 'em all a smack;
But don't mess with Muslims 'cos they may fight back.

I'm sorry if I've caused you all consternation;
But passion didn't die with my grandad's generation.
Whilst you all crow on about culture and tradition,
I've a friend on death row for changing his religion.

So take a walk outside your head;
Freedom is more than who's in your bed.
I want to reach out to the homeless, the poor;
The only way forwards is: 'Make Love not Law!'

So I'll hug the Christian, I'll call him my brother;
I'll pray with the Jews, they're my father, my mother,
I'll befriend the Buddhist and put out that fire;
And I'll love the Muslim … less you call me a liar.

For Daniel Scott

21ST CENTURY LOVERBOY

It's great nowadays, our thoughts are all free;
There's plenty of stuff on the net and TV;
Britons won't be slaves, we despise everything
as we turn a blind eye to every song that we sing;
We only have to believe what the scientist proves;
It means we can have sex with everything that moves.

Me mum went to church to make herself good,
and she tried to be nice whenever she could,
but now we ignore the Almighty's rod'
I mean, Dan Brown sold more books than God;
I'd gladly bow down and kiss the devil's hooves,
if it means having sex with everything that moves.

So come, join together this once great nation,
and fight for the right to see all sorts of penetration;
A curse upon love, trust and wedding rings'
People aren't human, we're all just things;
I hope when my kids grow up the situation improves'
I want them to have sex with everything that moves.

I LOVE YOU WITH ALL MY MADNESS

I LOVE YOU WITH ALL MY MADNESS,
I love you with all my breath,
I love you with tins of pencils,
I love you with all my death.

I wish I could paint you olive,
Hang velvet on your cross,
I'd like to start a barber's shop,
where you would be the boss.

I'd like to make a noise machine,
I'd like to call you 'Sir',
I'd like to fetch you water,
Like Charlton Heston in Ben Hur.

I feel like setting fire,
To every empty, no-where box,
I feel like putting itching-powder,
In Mephistopheles' socks.

I love you with all my madness,
I love you with all my paint,
I love you with every concept'
That makes my mother faint.

Do you get high on boredom?
Do you like making ants?
Do you really care what Tory Leaders
Keep in their underpants?

I love you with all my madness,
I love you with all my doubt,
I love you with every single thought,
That keeps the darkness out.

MARTIN COLLINS

Martin Collins was born in Worthing in 1954 and has lived all his life in Sussex. He currently works for a design and publishing company in Worthing.

EASTER

Is it eggs?
Is it bunnies?
Tulips or daffs?
Simnel cake,
Gardening,
DIY gaffes?

No, it's God in a garden
Making choices for me
Betrayed and deserted
Then nailed to a tree

It's Eternity
Yielding to time for a while
'Til death learns his place
Beneath Life's generous Smile

It's a wild celebration
Of Freedom and Grace
As justice and judgement
God's Mercy embrace

So, bring on the bunnies
The chocolate and flowers
This season of Life
Makes Eternity ours

WILL I STILL GET A PENSION?

When I'm old will I still get a pension?
I'll need one in order to live
My advisor neglected to mention
That in order to get, I must give

But I worked and I earned without giving
And spent all I earned willy-nilly
Now I find that I can't go on living
Without money; I feel rather silly

So I'm selling my house to raise money
And I'll live on the beach in a shed
Surviving on yoghurt and honey
'Til my money's all gone and I'm dead

I wrote this during a tedious company pensions seminar at work. My apologies to the nice man from Baird's Independent Financial Advisors. He did give sound advice, but this is about anyone who hasn't heard it!

THIS LOVE

THIS LOVE
This white-hot flame
Fuelled by infinite energy
Perpetually the same
Yet sparkles with diversity

This love
This treasured gift
Increases with each giving
Its value somehow lifts
As the giver is receiving

This love
This mutual care
Remotely need perceiving
An extra sense, so rare
Across all distance cleaving

This love
This immortal soul
A delightful fusion forming
Makes two halves one whole
Each to the other calling

This love
So long neglected
Yearns for its consummation
To see its hope perfected
Complete, one soul, one passion

This poem is about unconditional love. It could be about a love between two people or between God and man. However the reader sees it is right.

When I was growing up we had a big kitchen in which the whole family ate meals. Dad washed and shaved at the sink and us kids had baths in the sink until we were too big. We played wonderful games in the kitchen and it was the heart of our home.

When I was 12 my parents bought a nursing home and we had no family lounge or living room. All eating, TV watching, homework, discussing (arguing!) was carried out in the kitchen. It was a retreat for an angst-ridden boy trying to escape the barmy residents. As I grew and the nursing home was sold we moved to a house in High Salvington that also had a big kitchen.

All of my memories, from a toddler up until I left home, revolve around the kitchen. This warm and friendly room has always been the most special place for me and I would still rather stand and talk in the kitchen of a home than sit in the formality of a lounge.

It is in kitchens that I have found comfort, friendship, love, laughter, security, freedom and happiness; I still do.

I wanted 'The Kitchen' to be a simple poem, easy to read and quite nostalgic. So, the rhymes are fairly clean and the style deliberately naïve. It's an affectionate poem about my favourite room in the house.

THE KITCHEN

IT'S AN EARTHY PLACE, THE KITCHEN
Where life appears to settle
Food and family mixing
With cooker, cups and kettle

Dad shaves at the sink every morning
And the kids bath in it at night
After skating in socks and then falling
On the polished cork tiles for a fight

The floor, from an ice-rink soon changes
To a battlefield, muddy and grim
Where a toy-soldier conflict now rages
While dad scrubs his hands clean with Vim

Once a week the kitchen turned wash-house
When the twin-tub was loaded with clothes
Then I wished that I lived in a posh house
Whose laundry got sent down the road

For I hated the air damp and steamy
Our shirts hoisted up on the creel
The spinner vibrating and screaming
Expressing the sadness I feel

On Sundays the kitchen's a café
Our neighbours come over to drink
No Moccha, Cappuccino or Latté
Just instant with hot full cream milk

Of course, I remember mum cooking
Pancakes, hot curries, soused fish
I spent many hours just looking
To see how she created each dish

After spending a Saturday fishing
We'd first make a raid on the fridge
For lemonade, cheese or cold chicken
To satisfy three hungry kids

Yes, it's an earthy place, the kitchen
Where it's safe to express how you feel
To laugh, cry and argue or listen
To a family at home being real

MY VERY DEAR FRIEND

My very dear friend is now fifty
I've known him for most of those years
In body he's still just as nifty
But there's hair growing out of his ears

He hasn't gone all fat and flabby
His work on the land sees to that
But his hair is a little bit thinner
In the sun he must now wear a hat

Oh where do you go to dear David
When you're alone in your shed?
And the sawdust that lies all around you
Has mostly come out of your head

One summer we stayed in New Milton
In a tent on an old aerodrome
I rode on the back of your scooter
And fell asleep on the way home

I remember the back streets of Worthing
Two young men on a pub crawl
Both carrying their weekly pay packet
Leaving most of it up against a wall

Oh where do you go to dear David
When you're alone in the woods?
You can name all the trees that surround you
As every good scout leader should

We had lots of fun in our twenties
In bars, clubs and discos we'd roam
Now we've given up trying to be trendy
Weekends can be fun staying home

They say there's no smoke without fire
And fire Dave knows lots about
By rubbing two boy scouts together
The Raeburn will never go out

So where do you go to dear David
When you're alone on your bike?
Do you dream of the day we went fishing
In the pond and you caught a pike?

We talk quite a lot about Jesus
On every occasion we can
Unusual for two hefty geezers
We clearly both love the same Man!

Well, enjoy being fifty and healthy
Honoured by family and friends
You may never be very wealthy
But our love and respect never ends

So look into my face David Lewis
And remember just who we are
Whatever the distance between us
Our friendship shall never be marred

MAYBE TIME

TIME -
That frustrating thief of opportunity
Has conspired with cruel circumstance
To rob dreamers of their happy hopes
And the joyful consummation
Of pure, unconditional love

What little Time they have
Is like moments of joy
Laughter and passion
Distilled and condensed
So that, if unpacked,
They would fill the echoing atrium of eternity
Overflowing to return in a tumbling cycle
Of laughter and love
On and on and on…

Has Time slipped?
Have the cogs, levers and springs
Of the eternal movement
Somehow become unsynchronised?
Like blind dates, anxious to meet
Yet waiting under the wrong clocks
At different stations
They committed themselves
To two passing, kind and needy people
Not conscious of what potential awaited
If the merging of their souls had taken place

And so their lives continue.
Paths in time
Progressing along opposite sides
Of an aching chasm

Seeing each other
Tantalisingly close
Sometimes even touching
Briefly
As they venture out above the void
Along flimsy bridges built of dreams
Only to hurry back to their imposed pathways
As they watch another bridge of dreams
Spiral down into the cruel and hungry abyss
Of wasted Time

But wait!
Looking ahead they see their separate paths
Converging in the distance
Surely, if they just continue they will touch
And meet in an explosion of joy
At that precious point of future Time
Or is perspective painfully pointing out
Their illusion of consummation?

Maybe.
But maybe not.
Maybe Time is on their side after all
And all they need to do is wait
Long enough for him to realise his folly
To mend the rift he caused
By some distant temporal aberration
Then He who made Time
Made them
Made their love
Will finally give them His blessing
And join them together
For all Time
Maybe

Inspired by Romeo & Juliet, Dr Zhivago or any other star-crossed lovers

OPPOSITES

I SEE HIM MOST DAYS
As he stands, sentry-like
At his front door
Temporarily excluded like a leper
To breathe deeply of nicotine and tar
The man opposite

We did speak once
My car lights were on
And he knocked my door to tell me
But, normally we just wave
Silently watching out for each other
The man opposite

We swap cards at Christmas
The lady of his house writes those
She thanks me gushingly
For the excess apples we left in a box
On the pavement last autumn
The woman opposite

The lives of two households
Touching briefly, nervously
Like novice lovers
Exploring, but not taking risks
Or exposing tender roots
The people opposite

But these tentative touches show
They would care deeply, if needed
Spurred on by embryonic friendship
The smoking sentry would rush to help
Were more than careless car-lights
Burning opposite

LETTERS OF LIFE

Life is often merely letters
Distributed casually, thoughtlessly
Then mingled, once again ordered
Creating words, yelling marvellous meaning
Truly wonderful poetry
Yet making metaphors comprehensible
Expressing intimacy in every word
Forging and closing the ring of eternity

LOVE IS

Love is no parsimonious procrastinator,
Withholding its graces for fear of personal loss or hardship.
No, it empties itself lavishly again and again,
As if to immerse its object in luscious luxury.

ONE SOUL

THERE IS A LOVE THAT WORDS CANNOT MASTER
It defies expression by mere language
The communication of such a joining of souls
Can only ever be achieved
By gentle touch
Warm caress
Meeting of eyes
Understanding of hearts
Two lives surrendered
Each to the fulfilment
Of the other's dreams
Happiness, purpose, hope
For in such a consummation
There is no difference
And the hopes and dreams of two lives
Become the heartbeat of one.

For Lucy and Joanna on their wedding days.

FOR CHRISTOPHER

Why should I have to bury my son?
My burning tears can't purge the pain
that's caused by facing this – again.

For the second time
in eight short years
we gather, graveside once more,
to let our tears
wash the hallowed ground
that hides such fears

Why should I have to bury my son?
They say that time will soften the grief
ripping aching holes in what's left of peace.

Earth sprinkled softly
in drizzling gloom
falls gently as twin joins twin
in earthen womb
And brothers lie together once more
to rest in Heavenly mansion's room

Why should I have to bury my son?
We turn away to leave that place
and feel the sobbing comfort of friends' embrace.

There is no answer
to satisfy
as parents cry,
"Why did they die?"
But faith and hope shall stronger rise
And gentle strength soothe weeping eyes

Why should I have to bury my son?
Unsure, and full of fear and grief,
I search for God's elusive peace.

No easy answer
unravels the mess
that tangles my thoughts
and dreams, unless
I see beyond the angry, "Why?"
And trust the One who also died

I don't know why I had to bury my son
But I do know Him who holds the secret things
By whom all life expires – and begins

And I know that death has learned his place
beneath the feet of eternal Grace.
And twins now dance the song of life
beyond the reach of pain and strife.
Their tears are dried, as ours shall be
when we, with them, our God shall see.

For the family of Christopher Whiting, who died in January 2009 and went to join his twin brother, Jonathan, who died in 2000. Both absent from the body, but present with the Lord.